THE BIOLOGY, MANAGEMENT, AQUACULTURE AND FISHERIES OF LOBSTERS

THE BIOLOGY, MANAGEMENT, AQUACULTURE AND FISHERIES OF LOBSTERS

Edited by

Dr. Sandip S. Markad

Delve
PUBLISHING

www.delvepublishing.com

The Biology, Management, Aquaculture and Fisheries of Lobsters

Dr. Sandip S. Markad

Delve Publishing
224 Shoreacres Road
Burlington, ON L7L 2H2
Canada
www.delvepublishing.com
Email: orders@arclereducation.com

ISBN: 978-1-77469-839-6 (Hardcover)

ABOUT THE EDITOR

Dr. Sandip Markad is currently working as an Assistant Professor in Department of Fishery Science, Toshniwal Arts, commerce and Science college, Sengaon, Dist: Hingoli, Maharashtra. He completed his Bachelor of Fishery Science (B. F. Sc) in 2007 and Master of Fishery Science (M. F. Sc.) in 2009 from Dr. Balasaheb Sawant Konkan Krishi Vidyapeeth, Dapoli (MS). He qualified NET (Fishery Science) in 2012 conducted by National Agriculture Recruitment Board (ICAR), New Delhi and secured his Ph. D. in Fishery Science from Swami Ramanand Teerth Marathwada University, Nanded, Maharashtra in 2022. His specialization is Fisheries resource management and extension Education. He worked as research fellow in research projects during 2009 to 2013 and joined duties as an Assistant Professor from December 2013. He has published 20 research articles, 02 book chapters, 08 popular extension articles. He has completed two research projects sanctioned by Rajiv Gandhi Science and Technology Commission, Mumbai. He played key role of Organizing secretary for organization of one day National Conference on "Advances In Fisheries, Biological And Allied Research (AFBAR-2019)" during March 2019. He has organized 04 training programs on fish seed production and aquaculture techniques for fishermen and farmers, 02 Faculty Development Programs (FDP) on Educational Video Creation-Econtent Development for teaching faculties across the country. He contributed as resource person in 10 FDP conducted by different institutions in India. He participated in different conferences and workshops as well as presented his research work with one Best Oral presentation award. He has developed e-content through YouTube channel (Servavid) as well as institutional website. He is Permanent Active Member (PAM) of Asian Fisheries Society.

TABLE OF CONTENTS

LIST OF FIGURES

PREFACE

Lobsters have been the focus of research studies due to heavy fishing pressure for more than a century. There has been a steady increase in interest in the basic and applied biology of lobsters. This is due to the significant commercial importance of the lobsters as well as the fact that lobsters make excellent and fruitful research subjects in the fields of physiology, biochemistry and neurobiology. Despite having a wide range of species and a variable morphology, lobsters share many physiological traits and appear to have few significant ecological differences. A significant scientific advancement will improve knowledge of how lobsters succeed biologically. Their value as a fishery increases interest and broadens engagement as well as offering some guidance and support for parts of the studies.

The fact that they are large, many species with significant ecological functions makes them especially noteworthy. Almost all of the species are under heavy and uniform fishing pressure. The management of the precious lobster populations and the related research have received a lot of interest. The long history of the study of lobster biology serves as a good example of how basic and practical research are intertwined. Many lobster fisheries across the world are overfished as well as overcapitalized and their management plans are based on knowledge that is more than 50 years old. However, there are other requirements for good management in addition to biological understanding. The management of extensively fished lobster populations appears to be a political issue in many nations, in addition to being a biological and economic one.

For undergraduates and graduate students as well as research scientists who want to learn the fundamental elements of lobster aquaculture, this book will serve as a guide.

CHAPTER 1

Introduction to Lobster Aquaculture

CONTENTS

1.1 INTRODUCTION

Many countries harvest lobsters, which are ecologically and financially vital marine crustacean resource. It makes major contribution to the economies of at least few countries. Due to their commercial relevance as well as expanding academic interest in the fundamental and applied biology of this distinct group of crustaceans, few invertebrates have gotten as much attention as lobsters. Lobsters have been the topic of physiological, biochemical and molecular research over the past seven decades, with more research publications on lobster biology, ecology, physiology, fisheries and aquaculture (Radhakrishnan, Kizhakudan, & Phillips, 2019). In fact, *Homarus americanus*, the clawed American lobster, was the most studied species. The beginning of the twenty-first century saw a rise in molecular research, which has reconstructed the evolutionary relationships among the many lobster groupings. The lobster species namely spiny or rock lobsters, slipper lobsters, reef lobsters and blind lobsters belongs to the Suborder Macrura Reptantia of Order Decapoda under the Class Malacostraca of Subphylum Crustacea included in Phylum Arthropoda. Extant sea lobsters are classified into six families and 54 genera, with 260 species.

Figure 1.1. Clawed lobsters.

Source: https://cdn.shortpixel.ai/spai/w_750+q_lossy+ret_img+to_webp/https://wildernessclassroom.org/wp-content/uploads/2013/12/1024px-Fish4667_-_Flickr_-_NOAA_Photo_Library.jpg

The clawed lobsters (Figure 1.1) are related to the reef lobsters and the three groups of freshwater crayfish are sister species (De Grave et al., 2009).

Spiny lobsters are significant predators and prey on sharks, finfish and other marine species, as well as a keystone predator for a varied assemblage of benthic species. Due to the great demand for this unique seafood, indiscriminate and unregulated fishing has resulted in supply collapse in some nations. Due to the application of strong regulatory procedures, some well-managed fisheries exist around the world and these fisheries became sustainable.

1.2 LOBSTER

Lobsters are marine animals that are poikilosmotic over their salinity tolerance range, with only a few entering brackish waters. The mud spiny lobster *P. polyphagus*, for example, can survive a wide range of salinity. Lobsters can stay outside water for a long time if their gills are suitably hydrated and this trait is useful for live trade of lobsters over greater distsnces. Some lobster species are solitary by nature, whereas others are social. *H. gammarus* prefers stony substrates, but it can also dig into mud or create depressions in sand. *N. norvegicus* spends most of its life in burrows dug into muddy soils. Some of the species grow to be quite huge. The American lobster, *H. americanus*, grows to be 64 cm long and weighs more than 20 kg, whereas the Southern rock lobster, *Sagmariasus (Jasus) verreauxi*, grows to be 60 cm long and weighs more than 20 kg(Crear, Thomas, Hart, & Carter, 2000). The Scyllaridae family includes the tiniest lobsters. The entire body length of an adult *Eduarctus (Scyllarus) martensii* specimen is around 2.5 cm. The global lobster market is worth approximately $4 billion each year, with the Asia-Pacific area accounting for roughly 75% of total revenue. Only three families of marine lobsters namely, the Nephropidae (clawed lobsters), Palinuridae (spiny or rock lobsters) and Scyllaridae (slipper lobsters) are commercially important. Homarus, Nephrops, Metanephrops and Nephrospsis are the four genera that underpin the clawed lobster fishery. In 2015, the American lobster (*H. americanus*) accounted for 49% of total global output, with the Norwegian lobster (*N. norvegicus*) accounting for 15%, the Caribbean spiny lobster (*P. argus)* accounting for 12% and others accounting for the remaining 24% of total global lobster production (Butler, 2017). Palinurid lobsters are commonly referred to as 'rock lobster', 'spiny lobster', or 'crayfishes,' according to local or regional trade requirements. They sustain lucrative seafood trade and exports in many countries. 34 species covered under six genera (Jasus, Projasus, Sagmariasus, Panulirus, Palinurus and Puerulus) contribute to lobster fisheries.

Lobster species belonging to the genus Panulirus, provide the most extensive fisheries. Scyllarid lobsters are also known as 'slipper', shovel-nosed', 'sand', and they can be found all across the world's tropical as well as temperate oceans. Except for the few species that are useful in the aquarium trade, they are mostly caught as bycatch in commercial trawl fisheries. In some regions, they are occasionally used as a subsistence gillnet fishery. Thenus, Scyllarides and Ibacus are the most commercially valuable scyllarid genera. The larval development of nephropid lobsters is shortened (Figure 1.2).

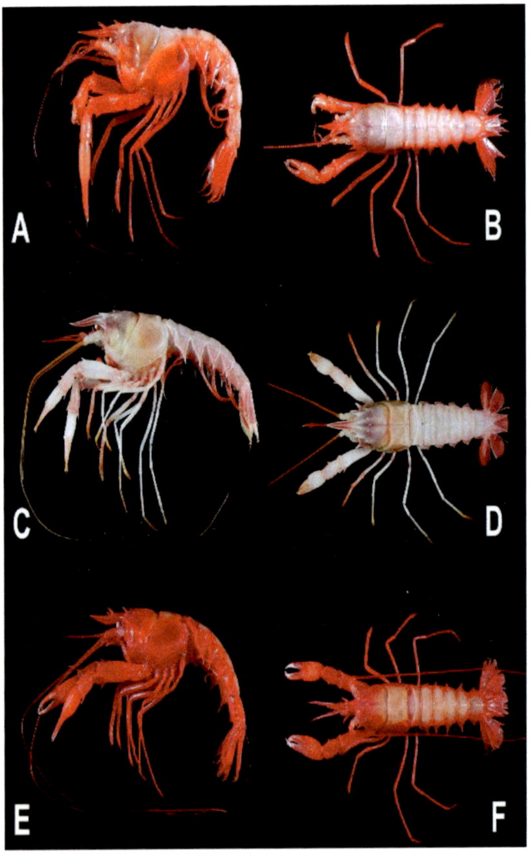

Figure 1.2. Nephropid lobsters (A-F).

Source: https://zookeys.pensoft.net/showimg.php?filename=oo_276852.jpg

The pre larva is followed by three larval stages and fourth post larvae stage, which is planktonic and quickly settling at the bottom before switching to a benthic habitat. Within three weeks, the larval stage is accomplished.

Spiny and slipper lobsters, on the other hand, have a protracted larval stage that lasts several months. The life cycle of these lobsters go through five stages *viz.* adult, egg, larva (phyllosoma), postlarva (puerulus or nisto) and juvenile. The juvenile phase has lately been separated into an early benthic phase and an older juvenile phase, both of which occupy adult habitats that are almost identical. The sexes are typically separated and some are hermaphrodites, although exhibit pseudohermaphroditism. Larval behavior and ecology of many species are still poorly known, despite the availability of relatively excellent information on recruitment processes and environmental influences of some shallow water species. Understanding the recruitment processes necessitates knowledge of larval dispersion patterns.

Lobsters are among the current marine arthropods with the largest bodies and the longest lives. Lobsters support commercially valuable fisheries in various parts of the world, including the most economically important in some areas. Understanding of growth processes is critical to understand the population dynamics, ecological function and sustainable management, since demographic activities like survival, reproduction and mobility are body-size dependent. The determination of age is a particularly difficult aspect of crab population research. The lack of obvious age markers in crustaceans makes a thorough understanding of the age-to-body size connection and the factors that influence its variability more important. Clawed and spiny lobster larvae have very different morphologies, behaviors and life spans. The size of larvae and post-larvae differed significantly among taxonomic groups. Larvae and settling post-larvae of clawed lobster are much smaller than the larvae and post-larvae of spiny and slipper lobster. It's crucial to understand the difference between stages and instars: the former refers to morphological changes, while the latter states the number of molts. There are just three physically distinct larval instars in clawed lobsters, stages I–III. There are significantly more larval instars in spiny lobsters, and the morphological phases of the phyllosoma are less reliably related to a specific instar. As a result, instars 13 to 17 of *Jasus edwardsii* completes all phyllosoma phases. Spiny and clawed lobster larvae are planktivorous, meaning they prefer to eat plankton (Jeffs et al., 2001). Lobsters metamorphose into a post larva, which is developmentally analogous to the megalopa in brachyuran crabs, after the final larval instar. This stage is known as the puerulus in spiny lobsters. The post-larval stage resembles the adult stage the most and is when the benthic existence begins. Unlike clawed lobster post-larvae, which continue to feed, the spiny lobster puerulus does not.

1.3 GROWTH

While the average growth factor between molts during larval development differs little between clawed, spiny and slipper lobsters, the greater number of larval instars in spiny and slipper lobsters results in dramatically greater overall length increase from first to last larval stage. The functional significance of the tremendous growth, endured by spiny and slipper lobsters during their larval stages, has received minimal discussion. Lobsters are at one extreme of the continuum, with indeterminate growth and the ability to reproduce at every instar after reaching adulthood. From few days in the larval and early juvenile stages to few years in large and older adults, the intermolt time rises with size. On the other hand, when body size increases, the percent increment of growth per molt normally decreases. Although the intermolt growth factor during the benthic stages is small in comparison to planktonic larvae, overall growth during benthic life is much greater, with the difference in length and body mass between the settling post-larva and sexually mature adult often being several orders of magnitude. Lobsters in their early benthic phase are often cryptic, sedentary and solitary, seldom leaving their nursery habitat. This habitat consists of a complex substratum like rocks, coral rubble, macroalgae or sea grass that protects the newborn lobster. Although differing levels of trophic specialization have been observed, most lobsters are omnivorous, eating mollusks, other crustaceans and algae. Suspension feeding has been observed in juvenile of *N. norvegicus* and *H. gammarus*. Juveniles go through an ontogenetic transition in which they either become more mobile or change habitats entirely (Perez Benavente, Uglem, Browne, & Marino Balsa, 2010). Spiny lobsters become more social and are typically found sharing shelters or moving in groups, whereas clawed lobsters remain primarily solitary in their shelter utilization.

Spiny as well as clawed lobsters grow at different rates and mature at different times. In general, taxa from warmer settings grow quicker and mature sooner than those from cooler climates (Dow, 1980). Males of all species physiologically mature at a smaller size than females, but it is believed that males must be as large as or larger than females to mate successfully. There is more than one mature instar and all female instars can become ovigerous after reaching maturity. The age and size at which the animal reaches maturity differs from one species to another. Maturation, in turn, slows down the development, with the effect being larger in females than in males due to the greater energy needed to reproduce. At maturity, taxonomically or ecologically comparable groupings can be quite different in size. Maturation also brings about allometric growth patterns that are sex-specific. Male lobsters reach adulthood earlier as well as at a lesser size than female lobsters and the instars during which they reach maturity are determined by the environment. When females reach maturity, their intermolt time lengthens in comparison to males of the same instar. Allometry differences between males and females can be good indicators of size at sexual maturity. In *H. americanus*, for example, males' claws grow larger while females' abdomens grow larger. In male spiny lobsters, the first and second pereiopods increase longer, but the surface area of pleopods decreases (Elner & Campbell, 1981).

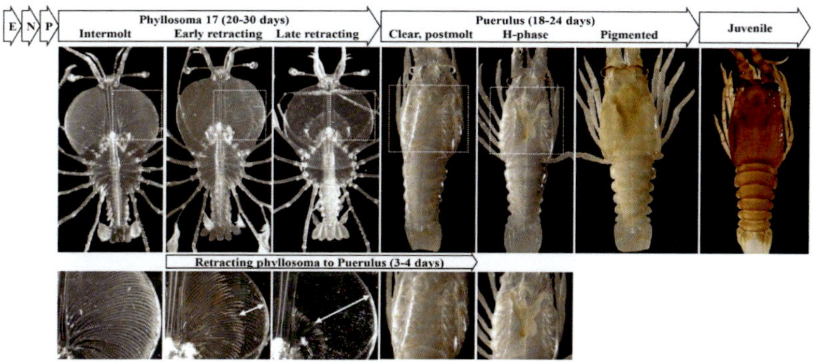

Figure 1.3. Molt cycle.

Source: https://media.springernature.com/full/springer-static/image/art%3A10.1038%2Fsrep13537/MediaObjects/41598_2015_Article_BFsrep13537_Fig1_HTML.jpg?as=webp

1.4 MOLTING

The major events of molt cycle are summarized in Figure 1.3. The old skeleton separates from the underlying epidermal cells during the molting process and a new cuticle is created, which thickens and hardens to form the new skeleton. Calcium carbonate is briefly retained in crystalline form as a pair of gastroliths on the lateral walls of the foregut as the old exoskeleton is decalcified from beneath (Hild, Marti, & Ziegler, 2008). Lobsters will ingest their cast-off exoskeleton once their mouthparts have hardened, if given the opportunity to conserve calcium. Stage A lasts for 24–48 hours after ecdysis, which involves the newly exposed soft and wrinkled integument stretching out to its now bigger size and depositing of the first of multiple inner layers of the new exoskeleton, the endocuticle. The endocuticle is located beneath the exocuticle and epicuticle, which were already in place prior to the molt. Stage B completes when the endocuticle's final layers are deposited. The exoskeleton achieves maximal rigidity during stage C as a result of chemical changes that harden the already deposited endocuticle, At this point, intermolt has occurred, a long period of time lasting until the commencement of physiological changes that prepare the integument for another molt. Stage D, also known as premoult or proecdysis, entails the separation of the old skeleton's endocuticle from the underlying epidermis, followed by the deposition of the new exoskeleton's outer layers viz. epicuticleand exocuticle. During this stage, demineralization causes a noticeable softening in portions of the old skeleton and the ecdysial sutures, which aids ecdysis. Water is swallowed and absorbed during stage E of ecdysis, increasing hydrostatic pressure within the organism. This causes the ecdysial sutures to break, such as the one along the dorsal midline of the carapace. The animal rolls on its side after 10–20 minutes of immobility, the exoskeletal membrane between the thorax and abdomen ruptures and the animal withdraws completing the cycle.

During the intermolt and early premoult stages, the majority of growth and regeneration occurs. It was found that muscle tissue grows in size through elongation, with the number as well as arrangement of muscle fibers remaining constant but increasing the quantity of thick and thin myofibrils. The muscles atrophy occurs temporarily during pre-ecdysis, just before the molt, due to enzyme-mediated degradation of actin myofibrils, presumably to help the animal withdraw from the old exoskeleton. This is likely to be accompanied by partial and temporary loss of mobility. Molt inhibiting hormone (MIH), associated crustacean hyperglycemic hormone (CHH) neuropeptides and terpenoid methyl farnesoate (MF) are among the hormones involved in molting (Nakatsuji, Lee, & Watson, 2009). Each of these hormone groups performs variety of roles, some of which change as it develops. The physiological changes that contribute to the molt are induced by molting hormones (ecdysteroids). The Y organ, a pair of hypodermal glands on either side of the thorax, produces this hormone family. Ecdysteroids titres in the hemolymph peak during premoult sub-stages D1 and D2, when the pre-exuvial cuticle is forming. When the old exoskeleton is being resorbed at substage D3, levels drop substantially and remain low after ecdysis from stage A to D0, when premoult commences.

1.5 HORMONES

The fundamental regulator of molting is the molt inhibiting hormone (MIH). It has a structural resemblance to crustacean hyperglycemic hormone

(CHH), which may similarly regulate molt. This family of neuropeptides is produced by the X organ–sinus gland complex, a specialized neuronal tissue found in the eyestalks. Increased levels of MIH, which are common during intermolt, prevent the Y organs from producing molting hormones. Environmental factors, particularly light levels, photoperiod and temperature, can influence the synthesis of MIH, explaining the molt cycle's response to changes in the environment, as the location of the X organ shows. MIH has a peptide sequence that is quite similar to CHH. However, one of its derivatives, CHHa, has both, a hypoglycemic and a molt inhibiting effect, whereas CHHb solely has a hypoglycemic effect. CHH levels appear to rise in the hemolymph at the end of the molt, possibly due to its involvement in boosting water retention during exuviation. Methyl farnesoate (MF), a sesquiterpene, is a precursor of juvenile hormone and can help regulate the molt cycle in larvae by delaying the molt. The mandibular organ is the source of MF. It is also hypothesized to play a role in reproduction in adult lobsters. Larger mandibular organs and greater hemolymph concentrations of MF in adult females indicate that it has a sex-specific function and may be crucial in sexual differentiation. Both direct and indirect methods have been used to measure crustacean growth. Rearing studies of caged animals or tag–recapture experiments in the wild have offered direct estimates of growth each molt or per unit period. The examination of size frequency distributions of samples of wild lobster populations provides indirect measurements of growth and size at age. Quantification of age pigments, metabolic by-products that increase with age, has been employed as a proxy for age in lobsters and crayfish with different degrees of effectiveness. Additional approaches, such as RNA: DNA ratios, can provide additional insight into the animal's nutritional health and, as a result, its growth potential.

Over the years, a considerable number of laboratory and hatchery-based studies of lobster growth have accumulated. Almost all larval growth research and a considerable number of post-settlement stage studies have been carried out in the lab. One of the important benefits of conducting laboratory growth studies is that it has provided a valuable setting in which factorial experiments can be conducted to evaluate heritable and environmental effects as well as detailed descriptive studies of the molt cycle. Temperature, photoperiod, space, substrate, feeding regimes and stocking density, among other variables, have all been studied through experiments. Many of these studies provide useful information on growth increment, intermolt duration and growth variables as well as demonstrating the possibility of highly varied growth patterns among individual lobsters.

In the field, precise estimates of intermolt duration are very difficult to come by. In this sense, laboratory observations have proven to be quite useful. However, caution should always be exercised when extending these findings to wild populations. The variance in the number of molts in younger age groups sets the scene for a wide range of lengths when it get older. It's crucial to note right away that, while carapace length is usually universally employed as a measure of lobster body size, it can't be considered to fluctuate isometrically with body mass. It's crucial to figure out the carapace-length-to-body-mass relationship for each species by sex, and preferably by area within each species. To our knowledge, no lobster taxa has undergone a similar comparison of carapace length to nominal length. In fact, the carapace length-to-body-mass relationship is one of the first morphometrics to measure in any research of growth, so that size-dependent processes, whether physiological or ecological, can be simply stated as a function of body mass.

Due to the broad availability of size data from population surveys, the examination of size-frequency distributions has been the most used indirect technique to measure lobster growth. When age classes already form fairly discrete modes in the size distribution, or when some prior knowledge of the age–size relationship and associated individual variability is available, this method works best. When it comes to long-lived crustaceans like lobsters, age classes overlap in size more often than not, obscuring age-specific features in the distribution. However, the younger age groups are frequently represented in the distribution as one or few prominent modes. While it is feasible to estimate the anticipated distributions of the different age groups merely by looking at the distribution, there are alternative approaches that provide a more statistically objective approach. MIX, ELEFAN and MULTIFAN are examples of statistical software (Fournier, Sibert, & Terceiro, 1991). Depending on which application is used, even these approaches require the user to make certain initial assumptions regarding the number of age classes in the population, the consistency of recruitment as well as the mean, standard deviation and skewness of the size distributions for each age class. These educated assumptions are then used to fine-tune the estimations via an iterative set of goodness-of-fit tests. Clearly, the validity of these assumptions has a significant impact on the quality of the study. MIX only considers a single length distribution, whereas the other programs allow for a time sequence of length composition estimations to be evaluated. Prior to conducting a size-frequency distribution analysis, the investigator should be aware of how sensitive the distributions can be to the sampling method. The investigator should be especially aware of the sampling gear's size-selectivity and how it might bias any representation of the size group. Trapping surveys, for example, are famously size selective, which may completely overlook early juvenile parts of the population that areotherwise well recorded by other methods at the same place. For example, ELEFAN can account for the effects of sampling technique selectivity.

1.6 POPULATION

The lack of physical age markers in crustaceans has made assessing growth in wild populations extremely difficult. Lipofuscin, the so-called age pigment, accumulates in the brains of crustaceans and has been shown to be a useful proxy for age in a variety of Crustacea, including lobsters (Medina, Vila, Megina, Sobrino, & Ramos, 2000). Lipofuscin is fluorescent under ultraviolet light, which allows for optical measurement. Since the density, sizes and shapes of pigment granules can be distinguished from any background fluorescence, fluorescence microscopy and image analysis of histological preparations of brain tissue became the preferred method over solvent extraction of the pigment. Initially, image analysis was done on paraffin-embedded olfactory lobe sections. Lipofuscin production is not, however, unaffected by environmental factors. It was found that its rate of accumulation increases with temperature, suggesting that it may be a better indicator of physiological age. While this may be conceivable where the thermal history of wild lobsters can be identified, it will be difficult for lobsters whose migration patterns result in significantly varied thermal regimes across their lifetime. As a result, in most situations, aspirations for the adoption of age pigments as a new standard for chronological age have yet to be realized. The procedure is nevertheless expensive and time-consuming. Nonetheless, these encouraging research findings justify a sustained effort to build an independent age measure.

Environmental heterogeneity is substantially responsible for regional and temporal variation in growth, development and sexual maturity size of lobster. Even among fish from the same mother, there is often substantial individual heterogeneity in growth, even when reared under comparable conditions. Small differences in start size have been demonstrated to be a major element in determining individual growth trajectories in communal rearing research with the American lobster. Differences in the thermal regime are one of the most commonly stated environmental factors, but others, such as food scarcity or intraspecific interactions, may also play a role. Stressful environmental situations can cause the molt to be delayed, growth increment to be slowed and even shrinking. Such environmental sensitivity in growth can have significant ramifications for population dynamics and management. We found no rigorous investigation of the degree to which heritable variation as a result of population diversification can explain regional differences in growth rate or maturity in the lobster literature. Despite their ability to control their position in the water column, pelagic stages are essentially at the whim of the thermal regime in which they drift during their initial weeks or months of existence. It may not be surprising that the majority of research on thermal impacts on nephropid lobster growth comes from a region where latitudinal and seasonal temperature variations are common.

Figure 1.4. American lobster.

Source: https://upload.wikimedia.org/wikipedia/commons/5/5e/Lobster.jpg

1.7 LARVA

The larval development of the American lobster (Figure 1.4) occurs over a wide temperature range. At temperatures below 10°C, however, it is impeded. From hatching to post larva, the development time varies from two months at 10°C to barely ten days at 22–24°C. Temperature influences larval growth as well. Larvae of American lobsters reared at 15–18°C were larger than those reared at higher or lower temperatures. Recent investigations in the Gulf of St. Lawrence support earlier predictions that larger larval and postlarval sizes are associated with warmer temperatures due to smaller growth increments associated with rapid development. However, geographic differences in the size of wild larvae do not always correlate with temperature, making it difficult to determine whether size-dependent mortality or the earlier settlement of larger individuals from the plankton are the cause of the differences in size. Because of the difficulty of keeping larval spiny lobsters and the lengthy period of many months required for larval development, there is less data on their growth. In general, larvae of cool water species like Jasus and Palinurus have been successfully reared at 20°C, while larvae of warm water species like Panulirus have been successfully reared at 25°C. The optimum temperature for growth in phyllosoma larvae of the subtropical Japanese spiny lobster, *P. japonicus*, is 26°C, although it swings downward to 24°C in more advanced larval stages (Fitzgibbon & Battaglene, 2012).

1.8 TEMPERATURE

Temperature differences in the growth rate of juvenile and adult benthic stages, especially in clawed lobsters, have been well documented and are easily linked to regional differences. Thermal stimulation of growth in the American lobster is related to temperature range of 8–25°C. Temperatures above 25°C are physically stressful or deadly, whereas temperatures below 5°C cause metabolism to slow to the point that the molt does not occur. Lobsters grow fastest in the southern Gulf of St Lawrence and southern New England, where summer temperatures exceed 20°C, but lobsters grow more slowly in the northern Gulf of St Lawrence and the Bay of Fundy, where summer temperatures are cooler. *H. americanus* has been grown in captivity to marketable size in two years at a constant temperature of 20°C. However, this species is unlikely to see that degree of growth in the wild. Regional or year-to-year temperature differences can significantly alter the timing and frequency of the molt in a lobster within the legal-size range, a factor with significant implications for annual recruitment to the fishery. Microwire-tagged European lobsters released at the age of one year in diverse thermal regimes in England, Scotland and Norway showed similar, but less significant thermal impacts on growth(Uglem, Belchier, & Svåsand, 2005).

Spiny lobster juveniles and adults are likewise affected by temperature, albeit they do not tolerate as wide a range of temperatures as clawed lobsters do. Warmwater forms grow quicker than cooler-water forms, with growth rates varying more widely among species. In general, most species grow faster in captivity if kept at somewhat warmer than ambient temperatures. For example, optimal temperatures for growth of *P. argus* and *J. edwardsii* are reported to be 29–30°C and 18–20°C, respectively (Thomas, Crear,

& Hart, 2000). However, raising animals at these temperatures can have negative consequences, such as reduced food conversion efficiency and an increased risk of sickness. Despite regional temperature changes, the growth rates of *J. edwardsii* are consistent along hundreds of kilometers of coast, covering a major percentage of the species' habitat. This could be explained in part by the fact that temperature fluctuations affect increment and molt frequency in different directions. For example, the general growth response of juvenile *J. edwardsii* reared under varied temperature regimes between 18 and 22°C, did not change in a growth experiment. This, however, conceals the reality that as molts became more frequent, the molt increment decreased (Marchese et al., 2019). Growth slowed at 24°C due to extended intermolt intervals, which could indicate physiologically stressful situations. The start of adulthood in lobsters is likewise accelerated by warm temperatures. Warmer-water lobsters not only develop faster, but they also mature at a lower size than their cooler-water counterparts. Female American lobsters, for example, mature between 68 and 76-mm CL in the warmer regions of their habitat, but attains maturity between 87 and 97 mm CL in the cooler areas. In New Zealand, the size at maturity of female *J. edwardsii* shifts from 72 mm to 122 mm CL as the temperature gradient decreases (McLeay, Doubell, & Linnane, 2019). The degree of claw and abdomen allometry in the American lobster is affected by thermally induced changes at the beginning of maturity.

The molt cycle can be drastically altered by manipulating the seasonal temperature cycle in the lab. In the spring, for example, if American lobsters are kept at 10°C or above, they immediately enter premoult and progress to ecdysis. The molt is inhibited even into the next summer if they are artificially kept at the same temperatures throughout the autumn. Furthermore, it was found that higher-than-normal winter temperatures of 15–20°C override seasonal molt inhibition, allowing the molt to occur at any time of year. Even yet, lobsters raised at the same temperature will have some seasonality in their growth cycle. In the absence of a seasonal temperature cycle, American lobsters reared from hatching at a constant 20°C temperature, showed a seasonal pattern of molting with the highest frequency in the spring as well as autumn and the lowest in the mid-summer as well as winter. This implies that factors other than temperature may influence seasonal patterns of growth. Lobster larvae grown in a culture with more food develop quicker, grow larger and survive longer, yet excess food can lead to less-than-ideal growing conditions. Lobsters, like many other crustaceans, require meals high in polyunsaturated fatty acids, carotenoids and astaxanthin pigments, which can be found in both algal and animal sources. Clawed and spiny lobster larvae are omnivorous, eating zooplankton and suspended particles. While clawed lobster post larvae are likewise planktivorous, the puerulus post larvae of spiny lobster species are not. In Homarus, larvae fed on other crustaceans either naturally occurring live zooplankton or commercially available brine shrimp grow quicker and are more robust than those fed with other meals. The growth of larval American lobsters is mainly independent of food supply as long as sufficient energy reserves have accumulated. Nevertheless, starved post larvae block the molt cycle in stage C and finally perish.

They discovered considerable changes in the percentage of poorly nourished post-larvae from year to year, while the implications of this variation for later development and survival were unknown. The transformation from a planktonic to a benthic existence was linked to post larvae having consistently greater levels of protein synthesis than newly-settled benthic lobsters. It was discovered that individuals of American lobster reared a few cm downstream from larger individuals were 40% smaller than those reared further downstream. The earliest juvenile stages of spiny lobsters are solitary, but they become social in their second year. As a result, growth rates did not differ between solitary and communally reared individuals in *P. cygnus*, but juveniles reared in isolation developed considerably slower after three years. Despite the favorable impacts of sociality on growth in most spiny lobsters, youngsters of *J. edwardsii* and *H. americanus* kept in captivity downstream of larger individuals grew slower. It was discovered that lobsters reared with both sexes together grew more slowly than those reared separately in *P. homarus* trials. The slowdown was linked to mixed-conditions of reproductive activity. The extent to which these impacts play out in natural populations is unknown.

Despite the differences in spiny and clawed lobster larval forms, the benthic stages of the various species show much more consistent growth and maturation patterns. The lack of a terminal molt allows lobsters to continue growing after sexual maturity and the various reproductive instars are all capable of generating eggs. From as tiny as 30 mm CL to greater than 100 mm CL and as early as 2–3 years to as old as 8–10 years, the size and age of maturity can vary greatly. Smaller and younger lobsters have higher proportional growth per molt than larger and older lobsters (Chubb, 2000). Near the commercially exploited size range, general patterns of incremental growth are well established, but little is known about the growth rates of early juveniles in nature. Because females invest more in reproduction, the arrival of adulthood slows their growth more than males. Sex-specific patterns of allometric growth of the claws, legs, pleopods and abdomen are also linked to the onset of adulthood. Because an individual's initial size and early growth can have a significant impact on their growth trajectory, additional long-term research on juvenile growth in the wild is needed. To assess the age–size relationship, a variety of techniques are available. We observed some of the most prevalent ways for determining lobster growth and age, both directly and indirectly. Tagging is likely the most extensively utilized direct method of obtaining wild stock growth data. However, workers should be mindful of the inherent biases that size-selective removal, tag loss, tag-related death and trauma can add to growth estimates. Given that size is not always an adequate proxy for age, determining age has been one of the most difficult tasks for scientists. In some species, physiological age indicators, such as lipofuscin, the 'age pigment,' can be used as a proxy for age. As previously stated, the procedure is costly, time-intensive and may not be widely applicable at this time. And, because lipofuscin accumulation is susceptible to environmental change, notably temperature, the index will need to be calibrated accordingly. Finally, it may be more advantageous for scientists to invest in a deep understanding of the geographical and temporal variability in the age–size connection at the population level. Monitoring programs that follow cohorts from the time of larval settlement and developing an understanding of the processes that lead to the observed size composition would be very helpful in this regard.

1.9 VARIABILITY

Variability in the environment has a significant impact on growth. Temperature is considered to be the single most important environmental element influencing lobster growth regionally and across the time. Not unexpectedly, the dominant influence of temperature on development is most visible where thermal extremes are highest, such as in the western north Atlantic, where *H. americanus* is found. Spiny lobsters have had the most reports of trophic influences on growth.

The Cape rock lobster, *J. lalandii*, (Figure 1.5) is possibly the most extreme example of 'bottom up' control on population size and body growth, with El Nio-driven changes in primary productivity influencing the abundance of a key suspension feeding prey for the rock lobster. Spiny lobsters have been found to have higher negative density-dependent growth effects than clawed lobsters, possibly due to their habits. The only exception is the cryptic, deep-water *N. norvegicus*, (Figure 1.6) which can likewise develop at a slower rate at a large density.

Figure 1.5. Cape rock lobster.

Source: https://upload.wikimedia.org/wikipedia/commons/thumb/7/70/JasusLalandii.JPG/800px-JasusLalandii.JPG

Figure 1.6. *N. norvegicus.*

Source: https://upload.wikimedia.org/wikipedia/commons/thumb/b/b1/Nephrops_norvegicus.jpg/800px-Nephrops_norvegicus.jpg

While the survival value of sociality in spiny lobsters has been emphasized, food constraint may be one of its drawbacks. The heritable component of lobster growth variation is much less understood. Even among siblings raised under identical settings, there is a lot of diversity in growth rates. While regional differences in growth have been well documented in many places and may be substantially explained by environmental differences, we are aware of no research on lobsters that have systematically assessed how much of those differences are the heritable outcome of population differentiation. This absence is most likely owing to the difficulties of performing the multigenerational studies required to make such assessments.

The genetic structure of lobster populations combined with comprehensive growth studies may be able to assist disentangle these effects. We discussed a variety of growth models, many of which were created with lobster population dynamics in mind. There are two types of models: continuous growth models and molt process models. Although significant progress has been achieved in the creation and use of molt process models, continuous growth models have been most widely used. As a function of size, there is currently no widely acknowledged functional form for the molt increment component or the intermolt duration. Rather than pre-moult–post-moult interactions, we propose that the size increase component be represented by the link between molt increment and pre-moult size. According to current evidence, a linear function is often acceptable for this purpose. Wherever possible, the probability distribution of molt increments in various size classes should be reported. Because of methodological restrictions, determining the association between intermolt time and size is more challenging. However, data is accumulating for number of lobster species based on both field and laboratory studies (N. M. Ehrhardt, 2008). The functional form chosen is crucial since the ramifications of the model choice might be rather significant, especially if projections are made beyond the range of available data. We advocate using logistic regression to model molt probability as a function of size. This method takes into account the underlying error structure and is limited to biologically appropriate values. The development of stochastic growth models for lobsters has progressed significantly.

Several different probability distributions have been used to depict variation in the time spent in different size classes, allowing for the modeling of intrinsic variability in growth. Because of the fundamental molt–no molt structure and the resulting quantum change in the size of molted individuals, a natural representation of variability in growth emerges for modeling approaches in which molt probabilities are provided. Molt increment is a minor part of overall growth variability, but it is nonetheless significant. Attempts to incorporate the causes of variability in the molting process by integrating external variables in crab growth models have received relatively little attention to date. In this regard, the development of models based on degree-day formulations for measuring 'physiological' time is particularly promising. Temperature, condition, food supply and other parameters can all be easily incorporated into the functional connections stated above for the size increase and intermolt duration components. Attempts to directly incorporate these aspects into models should be commended. In the formulation of molt increment–size and intermolt duration–size associations, these characteristics might be regarded as covariates. The need of considering temporal patterns in the components of growth should also be emphasized. Finally, considerably more data on individual heterogeneity in growth patterns is needed. This type of knowledge has hitherto been limited to laboratory observations for lobster taxa. It is unknown whether there are 'rapid' and 'slow' growth forms in general, or these patterns are genetically controlled as previously stated. To date, all models of lobster growth assume that the growth pattern is solely a function of size, without taking into account previous growth increments or intermolt periods. Lobsters have not been studied to see if growth increment or time between molts is dependent on previous growth. Conditional expectancies for molting have been developed in other crustaceans based on prior history and similar approaches should be used in lobster species whenever practicable. An approach like this could lead to the development of individual-based models for lobster populations that take into account the range of variation in demographic factors. There is a need to add essential intrinsic (heritable) and environmental elements that generate growth variability into growth models, as well as to continue the search for an appropriate age proxy. The development of age pigment approaches should continue, but it is also vital to have a better understanding of the elements that lead to growth variability so that we can better mimic a cohort's size trajectory in terms of both mean and variance in size through time. For this, increased emphasis will be placed on long-term field studies that perform the challenging task of tracking the destiny of cohorts over time.

CHAPTER 2

Distribution of Lobsters

CONTENTS

2.1 INTRODUCTION

Marine lobsters are often huge and have high economic value. New species have been discovered on a regular basis in recent years, after the publishing of an annotated checklist on the world's marine lobsters. 15 species of marine lobsters have been described since the previous checklist. There are now 260 species of extant marine lobsters classified into six families and 54 genera (Luypaert, Hagan, McCarthy, & Poti, 2020). It's still unclear whether Macrura Reptantia (Figure 2.1) is a natural group.

Figure 2.1. Macrura Reptantia.

Source: https://www.animalesmarinos.net/wp-content/uploads/2018/11/01-7-3.jpg

For example, a recent combined morphology and molecular analysis claimed that lobsters are not monophyletic, whereas recent molecular phylogeny has consistently shown that Macrura Reptantia is monophyletic. The relationships among the four infraorders of lobsters are likewise far from established. Thus, within the suborder Macrura Reptantia, which contains four infraorders, the higher categorization of marine lobsters continues to follow Chan's method (Tshudy, 2013). If a taxon's original name differs from its current generic designation, the original name is in square brackets at the end of the name. Only valid specific and subspecific names are assigned type locations. Some taxa still have unresolved taxonomic and nomenclatural difficulties, as indicated in the previous checklist. Decapods are a diversified group of crustaceans belonging to the Phylum Arthropoda that can be found in a variety of depths and latitudes throughout the world's aquatic ecosystems and terrestrial habitats. For species delineation, traditional classic taxonomy relied heavily on an organism's physical characteristics. The interrelationships of reptant decapods' principal clades are resolved using an integrative taxonomic method that combines molecular techniques like gene sequencing and DNA barcoding with traditional morphological traits. In the twenty-first century, molecular analytical technologies are increasingly being used to outline the evolutionary interrelationships of groupings within the reptantian decapods.

2.2 LOBSTERS

Marine lobsters belongs to the Macrura Reptantia suborder, which has four suborders, six families, 54 genera and 260 extant species. Palinuridae and Scyllaridae, two closely related families, share significant physical characteristics, particularly their phyllosoma larvae. The lobster fauna of India is so rich that the oceans surrounding the Indian subcontinent are home to 38 species belonging to three infraorders (Astacidea, Achelata and Polychelida) and five families (Chang, Chan, & Kumar, 2020). The south coast of India is rich in species because of the different habitats and physical conditions that encourage the colonization of a wide range of species. Though palinurid lobsters and the scyllarid *Thenus unimaculatus* can be found virtually anywhere throughout the Indian coast. *Petrarctus jeppiari*, a new deepwater scyllarid species, was recently introduced to India's scyllarid biodiversity. Lobsters are the most expensive crustaceans, with a healthy market and high demand all over the world, bringing in foreign currency for the country. Palinurid lobsters are known by variety of names in different countries. Scyllarid lobsters, on the other hand, are known by a variety of names in different languages, including slipper lobsters, shovel-nosed lobsters, bugs, locust lobsters, butterfly lobsters and more. Traditional classification relies heavily on morphological similarities between animals, whereas molecular systematics provides a useful tool for studying evolutionary interrelationships between organisms. Under traditional classification, the suborder Macrura Reptantia includes the superfamilies Nephropoidea or clawed lobsters, Palinuroidea or spiny and slipper lobsters, Eryonoidea or blind lobsters and Glypheoidea or the living fossil. Phylogenetic research using molecular markers revealed major changes in lobster taxonomy.

The Scyllaridae family, which is most closely linked to the Palinuridae family of the infraorder Achelata, is highly distinctive in morphology, with dorso-short and flattened antennae in contrast to a cylindrical antenna bearing a long multi-articulated and whip-like flagellum. The family Scyllaridae was divided into four subfamilies namely, Arctidinae, Ibacinae, Scyllarinae and Theninae, which accounted for nearly all of the species known at the time. Scyllaridae presently comprises 19 genera and 92 species divided into four subfamilies (Yang, Bracken-Grissom, Kim, Crandall, & Chan, 2012). The lobsters of the Polychelidae family are commonly referred to as deep sea blind lobsters and there are 38 species worldwide. The systematics of this family is currently being studied, particularly at the genus level. A total of 260 species have been added to the list. Despite the fact that Holthuis's list of marine lobsters throughout the world contained practically all of the species known at the time, numerous new species have been discovered in the subsequent two decades. A few new records have also been added to India's marine lobster list, which now includes 38 species.

According to the new system, species from the families Thalassinidae and Callianassidae that belong to the infraorder Thalassinidea have been included in the list. However, these two families are removed from the suborder Macrura Reptantia. The pattern of distribution of economically important palinurid lobsters in the Indian Ocean was documented. Researchers examined plankton samples collected by various expeditions (IIOE and DANA) and research cruises off the east as well as west coasts of India, to attempt the assignment of various stages of phyllosoma larvae to the spiny and scyllarid lobsters found throughout the Indian Ocean. The phyllosoma larvae of few species of lobsters were discovered off the western coast of the Andaman Islands. Adults of same species from this location, however, are yet to be documented. The existence of these larvae in this area most likely suggests that, these species are available in these marine islands. *Scyllarus batei batei* has been included in the lobster fauna when adults were discovered in the Arabian Sea and the Bay of Bengal. The phyllosoma larvae of *P. wieneckii* have been found on the southwest coast of India, but the adults are yet to be discovered. *P. mossambicus* was discovered on the southwest coast of India. However, the species was renamed as *P. waguensis* later on (Radhakrishnan et al., 2019).

2.3 SPECIES

The detection of novel species, differentiation of cryptic species, revision of taxonomy, assessment of evolutionary relationships, phylogenetic analysis, phylogeography and stock delineation investigations in marine lobsters have all been made easier due to recent breakthroughs in molecular technologies. The use of molecular markers to resolve contradictory difficulties in marine lobster phylogeny and evolution was discussed widely. In several fields of research, mitochondrial genomes have been used as a primary genetic marker. For reference, the full mitochondrial genomes of numerous lobster species have been characterized thus far. It is important to have a better understanding of the feeding of planktonic phyllosoma larvae of lobsters in order to commercialize lobster mariculture. The use of conventional markers and high-throughput DNA sequencing technologies to analyze the food of larvae has been explained. Understanding population demography and connections are essential for good fishery management. Management units can be accurately established by identifying the scale of subpopulation structure displayed by a species using genetic analysis. DNA microsatellites are common type of markers used in genomic analysis to determine population genetic structure, linkage, paternity and relatedness.

2.4 DEVELOPMENT

Recent developments in molecular technologies have had a significant impact on phylogenetic studies of marine lobsters. Their phylogeny had been a source of contention over issues like monophyly, the acquisition or loss of a stridulating organ, the direction of evolution between shallow and deep seas, and so on. A monophyletic group is a taxon that forms a clade that includes an ancestral species and all of its descendants. Though the concept of the monophyletic position of marine lobsters changes among researchers, many believe that they do not form a monophyletic group and relationships between taxonomic ranks differ from Holthuis' scheme (Ma, Chan, & Chu, 2011). Phylogenetic research has produced widely disparate conclusions. According to the most recent phylogenetic research using molecular markers, lobsters forms a monophyletic group, and several Thalassinidea members such as the squat lobster, which were not considered as real lobsters, were eliminated. The major three families that make up the infraorder Achelata are Palinuridae or spiny lobsters, Scyllaridae or slipper lobsters and Synaxidae or furry or coral lobsters. They all have a common larval phase called phyllosoma and their initial pair of pereopods lack chelae. There is strong phenetic and genetic support for the Palinuridae and Scyllaridae divergences. According to the most recent evolutionary analysis of Achelata and Palinuridae, Synaxidae should be either a synonym of Palinuridae or a subfamily of Palinuroidea.

Palinuridae is a family of 12 spiny lobster taxa divided into two lineages: Stridentes and Silentes, based on acoustic stridulating organs (Johan C. Groeneveld, Gopal, George, & Matthee, 2007). *Panulirus, Palinurus, Palinustus, Puerulus, Linuparus, Justitia, Nupalirus, Palinurellus and Palibythus* are among the Stridentes' nine genera, whereas Silentes has only three genera: Jasus, Projasus and Sagmariasus. It was demonstrated that Silentes is more primitive, constituting the basal lineage to Stridentes, using a collection of nuclear molecular markers. Stridentes were likewise shown to be monophyletic, with the synamorphic stridulating organ appearing just once during their evolution. The flattened antennal flagellum of slipper lobsters of the Scyllaridae family had been attracting less attention, possibly due to the decreased number of commercially relevant species. Scyllarid lobsters can be found in coastal waters along the continental shelf and upper slope areas across the Equator, in low latitudes as well as moderate latitudes influenced by warm water currents. Large scyllarid species are commercially fished, whereas lesser species are prized for their taxonomic variety. Based on the carapace forms and morphology of the maxilliped, exopods and mandibular palp, four subfamilies were proposed: Arctidinae, Ibacinae, Scyllarinae and Theninae. Molecular markers were used to determine the evolutionary relationships within Scyllaridae and there are currently four subfamilies, 20 extant genera and 89 species (Ferran, Rebeca, Michael, & Paul, 2016).

2.5 EVOLUTION

The evolution of lobsters is postulated by two hypotheses. The first proposed that lobsters evolved from a deep-sea ancestral population to shallow-water taxa, while the second proposed the opposite. Recent genetic investigations in the Palinuridae and Scyllaridae families, however, backed up the theory of a shallow-water reef origin, which later dispersed to deeper reefs before adapting to typical deep-sea bottoms. The focal taxonomic unit of conservation biology is the species. Molecular methods such as DNA barcoding have been widely used in taxonomy and biodiversity research and have proven to be effective in distinguishing related species. These technologies are also used in the enormous and diverse group of decapod crustaceans. Researchers were able to overcome several taxonomic obstacles by comparing tiny segments of DNA, leading to the identification of new lobster species. The 5′-end region of the cytochrome c oxidase 1 gene (COI) is classified as a DNA barcode or a standardized tool for molecular taxonomy and identification. Using mitochondrial COI, morphological variants of

Panulirus longipes with partially overlapping ranges could be divided into two subspecies, *P. longipes longipes* and *P. longipes bispinosus*. Crustaceans also utilize other mitochondrial genes, such as 16S rDNA. Researchers used morphological and molecular methods to reassign five different species in the previous monotypic subfamily, Theninae, to the genus Thenus (Jeena et al., 2016).

2.6 MARKERS

Molecular markers could also reveal speciation in some commercially important spiny lobsters. Panulirus argus was sampled from places in the Caribbean Sea and Southwest Atlantic, and COI and 16S rRNA genes, as well as nuclear gene sequences, were utilized to determine its taxonomic rank, resulting in taxonomic ambiguity. Researchers hypothesized the existence of a species based on the genetic difference discovered between specimens from the two regions. A study of mitochondrial markers (16S and COI regions) revealed the existence of at least six species in morphologically similar Palinurus taxa, leading to the discovery of *Palinurus barbarae*. Molecular markers have also been proven to be beneficial in identifying lineages. *Panulirus penicillatus* has been shown to have two highly different allopatric lineages from the eastern Pacific as well as the central and western Pacific. Molecular data analysis revealed that *P. homarus homarus* and *P. homarus rubellus* evolved as different lineages in the Indo-West Pacific. Concatenated datasets of nuclear and mitochondrial genes were used to predict evolutionary relationships as well as taxonomy conflicts in the infraorder Achelata. For all 17 extant Metanephrops species, Chan et al. (2009) attempted phylogenetic reconstruction using nuclear and mitochondrial genes. The monophyly patterns and phylogenetic relationships of Palinurus were determined using mitochondrial as well as microsatellite markers. Palinuridae phylogenetic reconstruction was also done using nuclear protein coding gene sequences.

Despite its tiny size, the mitochondrial genomic fraction is an important component of eukaryotic DNA. Mitochondrial DNA, also known as mtDNA, mitochondrial genome, or mitogenome, is a fundamental genetic marker utilized in an extensive range of fields including population genetics, evolutionary biology, phylogenetics, phylogeography, biodiversity and molecular ecology. Mitochondrial genomes of many lobster species have been characterized and are available for reference on the NCBI GenBank. A deeper understanding of the planktonic predatory phyllosoma larvae diet of any lobster species is critical for commercialization. There is a better chance that the mortalities suffered during larviculture might be minimized if their diet was nutritionally modeled after the natural diet. Soft-bodied zooplanktons may be the main prey for mid- to late-stage phyllosoma larvae of spiny and slipper lobsters. This finding could be supported by a Molecular Diet Analysis using 18S rDNA. The relative quantity of prey in ambient water correlates with the diets of phyllosoma, which are known to be opportunistic carnivores. For the phyllosoma of *Panulirus japonicus* and *P. cygnus*, a DNA-based diet method has proven to be a promising technique for analyzing diets. For *Jasus edwardsii* and *Scyllarus* sp., a set of group-specific primers was created to amplify DNA fragments from specific loci of potential prey, but not from lobsters. To identify prey species from phyllosoma midgut glands, high-throughput DNA sequencing techniques such as 454 pyrosequencing are used.

Resource managers seeking to rehabilitate overfished fisheries will benefit greatly from the information on the stock structure. Differences in a species' stock structure can indicate the existence of several populations with diverse gene pools. Failure to recognize the underlying genetic structure is a concern for commercially exploited species since it could lead to overexploitation and depletion of localized subpopulations, resulting in genetic drain and eventual loss of adaptive ability. New genetic techniques are robust enough to identify very small genetic changes within species and to uncover previously unknown spatial genetic structures, providing crucial information for the management of exploited marine species in the future. The discovery of spatial genetic structure can have a big impact on stock management. This could indicate that adults and larvae are not mixed together, casting doubt on the idea of a single population and implying that regional regulation would be more appropriate. It also gives a clear picture of the fishery's genetic health and evolutionary potential. Small effective population sizes and corresponding decreases in genetic diversity are early warning signs of a fishery at risk of declining production. Knowing the genetic structure of new specimens from both domestic and international sources would aid fishery managers in evaluating their contribution.

2.7 ENZYMES

For mtDNA, new marker types such as allozymes, AFLPs, full mitogenomes, RAPDs and RFLPs are rapidly replacing traditional marker types. Although there have been publications on the use of RAPD fingerprinting in population genetics in a variety of crustaceans, there is less data on this marker in lobsters. This method was useful in distinguishing the tissues of American and European lobsters as well as comparing different lobster species. *Homarus americanus*, *Panulirus marginatus* and *Nephrops norvegicus* were assessed using allozymes. After 2004, two types of markers, microsatellites and mitochondrial DNA-based markers, controlled the majority of lobster research. Mitochondrial markers have number of characteristics that make them excellent for use in population genetic research for variety of reasons. Because of its high mutation rate, mitochondrial DNA can give indications concerning population history over short time periods. With nucleotide sequence analysis of mitochondrial CO genes, the cytochrome oxidase II (COII) gene was used to investigate the genetic connectedness of the lobster *Jasus tristani*. When the genetic diversity of *P. homarus* populations in southern Sri Lanka and south India was evaluated using the COI gene, no significant differences were found. In many lobster species, population structure studies using mtDNA markers have been attempted (Singh, et al., 2018).

One of the most effective groups of markers in genomic analysis for determining population structure, linkage, parentage and relatedness is DNA microsatellites. Highly variable small tandem DNA repeats called microsatellites can be created and used in genetic research. Their core-repeat units are short, with lengths varying from two to six base pairs. The di-, tri- and tetranucleotide repetitions are the most prevalent. Even within maritime species with excellent larval dispersal capacities, microsatellites are useful markers for detecting genetic subdivision. Its data can also be used to calculate effective population size and population size fluctuations. For the commercially significant spiny and slipper lobsters, polymorphic microsatellite loci have been created in order to do population genetic studies. Microsatellite markers were identified in *Panulirus interruptus* and *H. gammarus* using partial genomic DNA libraries as well as in *Jasus edwardsii*, *Panulirus ornatus*, *Panulirus argus*, *Panulirus guttatus*, *Panulirus echinatus*, *Panulirus homarus* and *Panulirus echinatus* using next-generation sequencing (NGS) techniques.

Information on spatiotemporal patterns of population differentiation is one of the key elements in connectivity models for optimizing the size and spacing of marine protected areas (MPAs). Microsatellite markers that are polymorphic (msatDNA) are thought to be more effective than mtDNA markers at resolving population structure for small spatial scales. The Caribbean spiny lobster (*Panulirus argus*), which is protected by marine protected areas (MPAs) on the Mesoamerican Barrier Reef, may have a subregional population structure. Understanding how the many components of ecological systems interact is crucial for management, that is aimed at achieving sustainability since ecological systems are very complex and dynamic. Despite vast studies on lobsters, little is known about how they contribute to ecological functioning. Large size, high abundance and major trophic roles are the main reasons why lobsters are regarded as ecologically significant animals. To explain fluctuations in lobster abundance, it is necessary to comprehend the structures and functions of the lobster stock within the ecosystem. Between 65°N and 60°S, the temperate, subtropical and tropical oceans are all home to lobsters, with a vast variety of species living there. While certain Panulirus species can be found on sandy or muddy substrates, many prefer the protection of rocky or coral reefs. Although they can be found on a variety of substrates from the intertidal zone to depths of 200–400 m, the species of the genus *Jasus* are most frequently found on stony reefs (Butler, et al., 2006). In all of the oceans, temperate, subtropical and tropical regions between the latitudes of 4°S to 40°N and at depths ranging from 0 to at least 800 m, slipper lobsters can be found as adults. They inhabit a variety of environments, from soft substrates like sand and mud to firmer gravel surfaces, to rocky and reef crevices, and can be found in coastal waters, continental shelves and continental slopes. The habitat preferences of mature Palinuridae individuals range from shallow intertidal surf zones to deep ocean depths.

In order to precisely forecast the distribution pattern of lobster resources and create more effective management methods for these fisheries, knowledge of the life history processes and habitat preferences of lobsters can be useful. In their natural habitat, lobsters live alongside a variety of other vertebrate and invertebrate species. The wildlife in the lobster fishing areas has reportedly included sponges, sea urchins, starfish, ornamental fish, various decapods, seagrass and seaweed. Given that some of them make up the preferred diet of lobsters, there has been a direct correlation between the number of lobsters in the fishing areas and their associated fauna as well as flora. The delicate biological balance of the reef ecosystem may be greatly impacted by bottom-set gillnet lobster artisanal fishing and ongoing removal of the keystone low trophic species. For lobsters in the family Palinuridae, two distinct distribution patterns were observed: (i) a high latitude circumpolar distribution and (ii) a low latitude circum equatorial distribution. *Jasus*, *Projasus* and *Palinurus* are the genera that make up the first category. The Southern Ocean is home to the *Jasus* and *Projasus* species, each of which is confined to a certain area. *Palinurus* is presently restricted to the Atlantic and the southwestern Indian Ocean, but it may originally have had a circumpolar distribution in the Northern Hemisphere. The other palinurids that have an equatorial distribution are found at depths ranging from shallow water to over 1500 m in an extensive belt that extends about 35° on either side of the equator. *Puerulus*, *Linuparus*, *Palinustus* and *Justitia* are the genera with deep water equatorial distribution. Only the genus *Panulirus* is classified as having an equatorial shallow-water range. In comparison to the nine species of *Puerulus*, members of the genus *Palinustus* inhabit shallower depths (80–400 m).

Nephropidae family cinstitutes of 32 species in the tropical belt and 15 in the subtropical zone. The number of species under the genera *Nephropsis* and *Metanephrops* are highest in the tropical region. The southern Atlantic and Indian Ocean regions are home to three species from the two genera *Thymops* and *Thymopides*. While the distribution of scyllarids is constrained between 45°N and 44°S, that of palinurids ranges from 60°N to 50°S. Out of 63 species of palinurids, 39 are tropical species and 19 are subtropical species, with their distributions mostly overlapping. The Southern Hemisphere's latitudes are the only places where *Jasus* is found: subtropical (*J. caveorum*, *J. frontalis*, and *J. lalandii*) and temperate (*J. edwardsii* and *J. paulensis*). All three of the world's major oceans include the palinurid species *Justitia longimanus*. Currently, 24 species and subspecies of the genus *Panulirus* are known and they are spread throughout the tropical and subtropical belt of all the oceans i.e., between 35°N and 35°S Latitudes, wherever suitable habitats are present for their existence in depths below about 100 m. Out of them, 8 species are found in subtropical latitudes while 16 are found in the tropical zone. There are 6 species in the genus *Palinurus* and 5 of them are found in temperate and subtropical climates. The Glypheidae family of living fossils consists of 2 tropical genera and 2 species. There are 12 species in the family Enoplometopidae, of which 9 are found in the tropical zone and belong to the single genus Enoplometopus. There are 38 species in the family Polychelidae, divided into 6 genera, with 29 of those species found in tropical latitudes.

With 131 species, the Indo-West Pacific has the highest species diversity. Nephropids are represented by 29 species, palinurids by 36 species and scyllarids by 57 species. Of the 24 species including 4 subspecies of *Panulirus*, 13 are found here (Lavalli, Spanier, & Grasso, 2007). Two species of the genus *Jasus* live in the temperate zone lower than 35° S, while the other 3 are found in subtropical latitudes. One species, *J. edwardsii*, is found in both the southern Indian Ocean (southern Australia) and the Pacific Ocean, while two of the five species are found in the Pacific Ocean (Southern New Zealand). The only species that lives in the Atlantic Ocean is *J. lalandii*. Both the southern Atlantic and Indian Oceans have *J. paulensis*. Though physically separated by thousands of miles, population genetic studies and analyses of molecular variation found no evidence of a significant population genetic difference between *J. tristani* and *J. paulensis*. Lobsters species of *Puerulus*-genus are often deepwater species. The Indo-West Pacific region is home to all 9 *Puerulus* species. All of the species that make up the genus *Linuparus* are found in the Indo-west Pacific. *Panulirus* has eight subtropical species, five of which are found in the Indo-Pacific. The SW Indian Ocean is home to one subtropical species of *Palinurus*. One of the two temperate species of *Palinurus (Palinurus gichristi*) is found in the southern Indian Ocean and *Palinurus elephas* is found in the North Atlantic Ocean.

There are 63 species of scyllarids spread in the tropical zone, 26 species in the subtropical zone and 3 species in the temperate zone. Scyllarids only occur between the latitudes of 45°N and 44°S, with the majority of taxa occurring within those latitudes. 16 genera and 57 species make up the Indo-West Pacific region; 13 of these species only occur in the Pacific Ocean, 22 in the Atlantic Ocean and 1 is found in all three oceans(Lavalli et al., 2019). The genus *Scammarctus* (*S. batei arabicus*) and *Petrarctus* (*P. jeppiari*), both having a single species, are endemic to the Indian Ocean. The former may be found in the Gulf of Aden, while the latter can be found in Muttom, Kanyakumari district, on the southwest coast of India. The subtropical species *E. princeps* is found in the Baja California region on Mexico's west coast. Six of the 8 species that make up the genus *Ibacus* are found in the Indo-West Pacific. While the eight species belonging to the genus *Scyllarus* are only found in the Atlantic and Mediterranean regions, the genera *Arctides*, *Acantharctus*, *Scyllarides* and *Bathyarctus* are represented in the Atlantic and Indo-West Pacific regions. The Indo-West Pacific region is the only place where the five species of the genus *Thenus* are found. The southern temperate zone has three species from two genera (*Antarctus* and *Ibacus*). The Indo-West Pacific region is home to three different *Parribacus* species. *H. americanus* and other macruran decapods are strongly connected with habitats that provide refuge until they outgrow their vulnerable size. Certain aquatic species exhibit ontogenetic habitat alterations to promote development and reduce predation danger. Many spiny lobster species exhibit ontogenetic habitat shift from the postlarval settlement habitat of macroalgae, kelp or seagrass to benthic crevice sheltering concomitant with aggregation in crevices as larger juveniles and subadults. Some species exhibit ontogenetic changes as they get bigger and bigger, even sharing a den with their own kind. *P. argus* juveniles raised underwent transition at a smaller size than lobsters raised alone under the same circumstances. Larger lobsters wait until they are big enough to switch from living in algae to hiding in crevices in the absence of conspecifics. Conspecifics' proximity is likely factor in the ontogenetic habitat alterations of animals.

2.8 DISTRIBUTION

In their natural environments, lobsters live alongside both vertebrate and invertebrate species. The southwest coast of Kanyakumari (India) shoreline includes coral reefs in the deeper waters and rocky outcrops in the shallow waters, which create the perfect environment for lobster colonies to grow. The spiny lobsters find favorable protection in the partially and completely submerged rocky outcrops with rocky crevices covered in oysters and macroalgae. The presence of the Brown mussel, *Perna indica*, which serves as the main food source for spiny lobsters, is linked to the high quantity of spiny lobsters along the Kanyakumari coast between the months of October to January. On the intertidal and subtidal rocks, *P. indica* makes up roughly 10.5% of the fauna. In the fishing grounds at Khadiyapatnam, a significant lobster fishing hub in the Kanyakumari district, associated fauna caught in bottom-set gillnets (BSGN) include 10 genera of fish, 5 species of crustaceans, 3 species of mollusks, 10 species of sponges, 3 genera of echinoderms, and 14 species of seaweed (Lenihan et al., 2021). The starfish and brittle stars are metaphors for echinoderms. While sponges and echinoderms predominate during the fishing season (October–January), the composition varies throughout the non-fishing months (April–May). Young lobsters are significantly protected by macroalgae, seagrass, sponges, gorgonians and rocky crevices, which lowers the risk of predation. An average of 70 gillnets are run per day during the three busiest fishing months at Khadiyapatnam, landing an estimated 25 ton of the accompanying wildlife and flora as bycatch.

In situ and *in vitro* investigations were conducted inside marine protected areas in Tasmania to demonstrate how the indiscriminate removal of large predators like *Jasus edwardsii* could significantly weaken the resistance of kelp beds against the soaring sea urchin population.

Figure 2.2. *Strongylocentrotus purpuratus.*

Source: https://upload.wikimedia.org/wikipedia/commons/thumb/f/f4/Strongylocentrotus_purpuratus_1.jpg/800px-Strongylocentrotus_purpuratus_1.jpg

Strongylocentrotus purpuratus (Figure 2.2) is preferred by the spiny lobster *P. interruptus* over *S. franciscanus* as a food source off the coast of southern California. Although they are capable of feeding on larger sizes, lobsters preferred smaller sizes when provided different sizes of *S. fransiscanus* in the lab, perhaps due to handling challenges. Large-scale lobster exploitation caused sea urchin populations to explode, which in turn led to outbreaks of destructive grazing of kelp forests in the 1950s. In light of the fast-rising climate and unprecedented elimination of predators from the ocean systems, it is important to establish more comprehensive resilience-based strategies. There may be a number of other factors outside predator removal that cause trophic cascades in an ecosystem. Nutrition availability, wave energy, sedimentation and the interaction of all these factors may also affect macroalgal abundance.

Enhancing aquaculture practices requires knowledge of the molt cycle. The life history of most crustaceans is coordinated with the molt cycle, which is a complex and energy-intensive process. The time between two molts, known as the intermolt, is when tissues are replaced by the water that was absorbed during the molting process. The intermolt period is when adult females finish their whole reproductive cycle, including mating, ovarian maturation, yolk synthesis, egg extrusion, fertilization, and parental care of the young. Molting and reproduction are thought to be influenced by variety of environmental, physiological, behavioral and dietary factors. These two metabolic processes are regulated by number of multifunctional neuroendocrine hormones situated in the brain system and mandibular areas. A better understanding of the interactions between molting and reproduction in crustaceans has been made possible by developments in crustacean endocrinology over the past 30 years, including the development of more reliable analytical methods to isolate as well as characterize the hormones and the identification of mRNA genes responsible for their production. Numerous biotic and abiotic factors affect the pace of growth in various species and related species in various geographic locations. Food is one of the biotic elements that has a significant impact on how quickly crustaceans grow. The growth rate of lobsters is determined by the food's quality, accessibility and density in proportion to food supplies. The primary abiotic element affecting the growth of temperate and semi tropical species is temperature. Salinity, oxygen and photoperiod are few other abiotic variables that have an impact on lobster growth and reproduction.

2.9 GROWTH

Growth is largely influenced by food. Most decapods are omnivorous, while some are only carnivorous and some exhibit signs of saprophagia. Studies on nutritional needs of animal, interactions with other species and potential for cultivation require knowledge about natural diet of the animal. According to some researchers, palinurid lobsters are benthic carnivores and have been described as omnivorous with diets that include crustaceans, gastropods, asteroid echinoderms, porifera and marine plants. Many aquatic animals, including sharks and octopuses, feast on them. Spiny lobsters display ontogenetic changes in how they utilize their habitat during the juvenile and adult phases for various behavioral goals exhibiting progressive nutritional shift as they get bigger. They have an impact on variety of prey species and habitats, play many roles including those of prey, predators and keystone species. Like other decapods, spiny lobsters periodically shed their old exoskeleton in order to grow or increase in length and weight. The chitinous lining of the esophagus, foregut and surface of the gills are also peeled off during ecdysis in addition to the exterior body shell. The flat, leaf-like phyllosoma larva completely changes into an adult-shaped puerulus at the end of the larval phase. Both homarid and palinurid lobsters go through the same ecdysis process, albeit the time required to complete each step vary. Several morphological, anatomical and physiological changes occur as the lobster enters the proecdysis or molt preparation phase (Cheng & Chang, 1991). Activities associated with proecdysis include selective ion and water absorption, limb regeneration, changes in metabolic pathways and resorption as well as storage of cuticular components in the haemolymph. Ecdysis has been separated into two periods for convenience: a passive phase and an active phase.

The carapace separates from the branchiostegal decalcified ecdysial sutures above the bases of the legs during the passive phase The thoracoabdominal membrane bulges outward at the junction of the carapace and the abdomen due to water absorption. The phase may be prolonged for several hours if the lobster is disturbed or if the circumstances are unfavorable. The thoracoabdominal membrane ruptures as the lobster enters the active ecdysial phase and the lobster either falls sideways or maintains its upright position upon losing the grasp while wriggling and jumping out of the previous exoskeleton. The lobster continues to take in a lot of water during the post ecdysial period, stretching its limbs and body to its new length and volume. The lobster begins actively feeding during the metecdysis phase, during which a new cuticle is formed, mineralization takes place and the deposition of new tissues begins to replace the water absorbed during ecdysis (Cheng & Chang, 1991)Homarus americanus, was further divided into five substages according to the degree of cuticle digestion along the dorsal midline of the carapace and on the dorsal surface of the merus of the chelipeds. The mean times to ecdysis for intact lobsters (5.5 +/- 1.1 g.

The lobster is incredibly adept at maintaining synchronized movement and agility. Until the rupture of the thoracoabdominal membrane and splitting of the carapace from the abdomen, the lobster is capable of maintaining a high level of coordinated activity, agility and neuromuscular control of the body functions, but becomes defenseless and vulnerable during the final phase. Newly molted lobsters in captivity avoid other lobsters by scurrying to safe locations.

The molt likelihood for shorter intermolt times was highest for smaller animals in both males and females. Larger animals typically have longer intermolt times as their growth continues. Normal lobsters immediately consume some of the shedded exoskeletons after ecdysis, which is likely a mechanism to replenish calcium and minerals lost through their exuviae. Unlike other species like *J. lalandii*, the captive spiny lobster *P. homarus* has never been associated with cannibalism. However, *P. polyphagus* showed a significant propensity to consume molt exuviae and recovery of full molt from *P. polyphagus* (Figure 2.3) reared in tanks was consistently challenging. However, *T. orientalis* did not show any preference for the exuviae of the molts, making it nearly always able to collect the molts. The newly molted lobsters hide in the refuge and stay away from other lobsters. On the other hand, if normal and eyestalk-ablated (EA) lobsters are maintained together, the normal lobsters will easily feast on the EA lobsters.

Figure 2.3. *P. polyphagus.*

Source: https://upload.wikimedia.org/wikipedia/commons/thumb/c/c0/Panulirus_polyphagus.jpg/800px-Panulirus_polyphagus.jpg

India and the nations bordering the Indian Ocean collect lobster, a rich crustacean resource. The export of lobster brings in an average of Rs. 196 crores (USD 32 million) in foreign cash each year, although making up only 0.05 percent of India's annual marine seafood catch (Stoll et al., 2018). The family Palinuridae and Scyllaridae are home to the commercially exploited lobsters in India. Both the artisanal and mechanized sectors heavily exploit the multispecies resources, with the latter accounting for roughly 67 percent of the entire catch. Only 4 species of palinurids (*Panulirus polyphagus*, *P. homarus homarus*, *P. ornatus* and *Puerulus sewelli*) and 1 species of scyllarid (*Thenus unimaculatus*) considerably contribute to the fisheries out of the 38 species that make up the lobster fauna of India.

Despite the fact that lobsters are found virtually along the entire length of the Indian coast, the largest fisheries are found in the northwest, northeast and southern parts of the country. The northeast state of West Bengal along with the two states of Gujarat and Maharashtra in the west together account for an average of 65% of the nation's annual catch. *P. polyphagus* is the predominant species throughout the northwest and northeast shores, making up 61 percent of the total yearly harvest. The three main species that make up the fisheries throughout the southern region are *P. homarus homarus*, *P. ornatus* and *T. unimaculatus* (Jones, 2007).

On India's southwest coast, the Kanyakumari district of Tamil Nadu was previously the primary area for landing lobster. The average catch, which peaked due to excessive exploitation and a lack of efficient fishing restrictions, gradually decreased later on. The three main species supporting fisheries in Kerala are *T. unimaculatus*, *P. homarus homarus*and *P. sewelli*. With sparse landings in recent years, deep-sea lobster fishing is concentrated in the southern region. According to stock assessment research on the *P. polyphagus* fishery in Mumbai, the estimated combined average MSY of males and females (168.42 t) is very low compared to the yield (243.5 t) and therefore a reduction of 40 percent effort is required to ensure the stock's sustainability. Due to the fact that the resource is a bycatch in trawlers, it is difficult to put fishing limitations into practice, which makes it difficult to manage the spiny and slipper lobster trawl fishery. Mesh restriction and a lobster fishing ban during the months when reproduction is at its highest are two management strategies that could be used in the artisanal fishery. The minimum legal size (MLS) for lobster export has been announced by the Central Government. However, state governments, with the exception of Kerala, have not enforced the MLS for fishing. By establishing Fishery Management Councils including representatives of fishermen, scientists, administrators, fishery experts and NGOs, co-management approaches to control artisanal fishing with gillnets are being examined. The resource might be saved from further decline if the fishermen can agree on a minimum legal size for fishing, a prohibition on the use of gillnet as well as trammel net for lobster fishing and decisions for fishing holidays during the peak mating season.

2.10 FISHERY MANAGEMENT

The unorganized lobster fishery management in India is explored in relation to the resource management and lobster fisheries in the nations bordering the Indian Ocean. Many nations around the world economically exploit lobsters, which forms a low-volume and high-value crustacean resource with considerable export demand. They are appealing to fishermen as a source of income and as a food source for people because of their high unit value, straightforward harvesting techniques and potential for live transport. Both the artisanal and industrial sectors actively utilize the resource and the majority of global stocks are either fully or even overexploited. Overfishing, habitat damage, climate change and coastal pollution are all potential contributing factors to further fall in landings.

CHAPTER 3

Management of Slipper Lobster

CONTENTS

3.1 INTRODUCTION

Research on many elements of lobster biology has grown significantly over the past 30 years, but most of this growth has been concentrated on the biology of clawed and spiny lobsters. Research on slipper lobsters is still in its infancy and to some extent, our present understanding of them is less than it was for clawed and spiny lobsters thirty years ago(Alboréset al., 2019). Simultaneously, slipper lobsters are increasingly being used as bait in commercial fisheries or as by-products in other fisheries. As a result, it is crucial to increase our biological understanding of these species in order to implement sustainable management plans that will ensure the survival of slipper lobster stocks.

Figure 3.1. Scyllarid lobster.

Source: https://upload.wikimedia.org/wikipedia/commons/thumb/7/71/Scyllarides_latus.jpg/800px-Scyllarides_latus.jpg

The scyllarid lobster's (Figure 3.1) body is segmented. The segments or somites are fused to form three distinct body regionsnamely, the cephalothorax made up of six cephalon somites fused with eight thoracic somites, the abdomenwith six separate somites and the telson, which is sometimes referred to as a seventh abdominal segment or a medial appendage of the sixth abdominal segment. Each body segment bears paired appendages that are highly specialized structures and are fundamentally biramous. The mouthparts are located on segments 4 to 6 of the head, along with the antennules, also known as the first and second antennae. The pereiopods are carried on segments 10 to 14, whereas highly modified thoracic mouthparts are carried on segments 7 to 9. On segments 15 to 19, the abdomen has paired pleopods that are used for forward swimming in nistos and for reproduction in adults (Lavalli & Spanier, 2007). The tail fan's modified pleopods or uropods, which make up the final abdominal segment, are present.

3.2 CHARACTERISTICS

Important taxonomic characteristics used to distinguish species often include the shape, pubescence and spination of the carapace, which covers the cephalothorax and may expand anteriorly to create a rostrum and laterally to the legs to enclose the branchial chamber. The size, form and dentition of the second antennae as well as the structure and length of the antennae, all aid in identifying different species. The antennule's bifurcated flagellum is carried by a long, incredibly flexible and three-segmented peduncle in scyllarids. The flagellum of scyllarids is reduced to a single platelike segment and is attached to five extra, highly flattened antennal segments, in contrast to the whip-like flagellum of the second antennae in nephropids and palinurids. Scyllarids stand out from the other lobster families alone due to this characteristic. There are three genera and fifteen species in the subfamily Ibacinae. In all species, the dorsoventral compression of the carapace is severe and the lateral margin of the carapace has a deep cervical incision. In contrast to the Arctidinae, the mandibular palp is either simple or two segmented. A posteromedian tooth, a closed cervical incision, completely closed orbits and smooth-tuberculate body are the distinguishing features of the one and only species of *Evibacus*. In contrast, the species of *Ibacus* and *Parribacus* have anteriorly open orbits and a conspicuous open cervical incision. The dorsal body surface is smooth and punctuated with pubescence in *Ibacus* but coarsely squamose-tuberculate in *Parribacus*. The fifth abdominal segment is spined in *Ibacus* but spineless in *Parribacus* and the number of mandibular palp segments are differences between the two species.

The subfamily Theninae was previously thought to be monospecific. This subfamily and the Scyllarinae share the lack of a flagellum on the exopods of first and third maxilliped. They differ in how far away from the anterolateral edge of the carapace their orbits are located in the Theninae against the Scyllarinae and how much of the body is compressed dorsoventrally in the Theninae versus the Scyllarinae. Fourteen genera, 49 recognized species and two currently unknown species make up the Scyllarinae subfamily. People have been aware of the exceedingly unusual family of slipper lobsters, the Scyllaridae. 85 scyllarid species from 20 genera and four subfamilies have been identified (Yang et al., 2014). It is inevitable that more species will be discovered, especially in the Scyllarinae subfamily, which comprises more than half of the scyllarid species and is dispersed over a wider range of latitudes and depths than the other three subfamilies. The Palinuridae and Synaxidae are closely related to Scyllarids. The three families are distinguished from other Decapoda by a number of characteristics, most notably their phyllosoma larvae.

Figure 3.2. Slipper lobsters.

Source: https://seaunseen.com/wp-content/uploads/2018/11/MG_7360.jpg

Slipper lobsters (Figure 3.2) may easily be distinguished by their antennal flagellum with a plate-like shape from palinurids and synaxids having whip-like antennae. Characters that clearly distinguish higher taxa from lower taxa include the presence of multiarticulate exopods on the maxillipeds of the subfamilies Ibacinae and Arctidinae as well as their absence from the Theninae and Scyllarinae. Less is known about the relationships between the four subfamilies and it is unclear if some physical traits are derived or primitive. The Scyllaridae are Tethyan in origin and have evolved along with the development of the major seas, according to fossil and morphological evidence, species distribution, developmental traits, ecology as well as biology. Scyllarids seem to have recently expanded into habitats in shallow, shelf and deep waters. The family is pantropical; thus, it is highly unlikely that any group within it, or the family as a whole, is a remnant of a population that was once more widespread. However, some distinctive genera are local to a particular area.

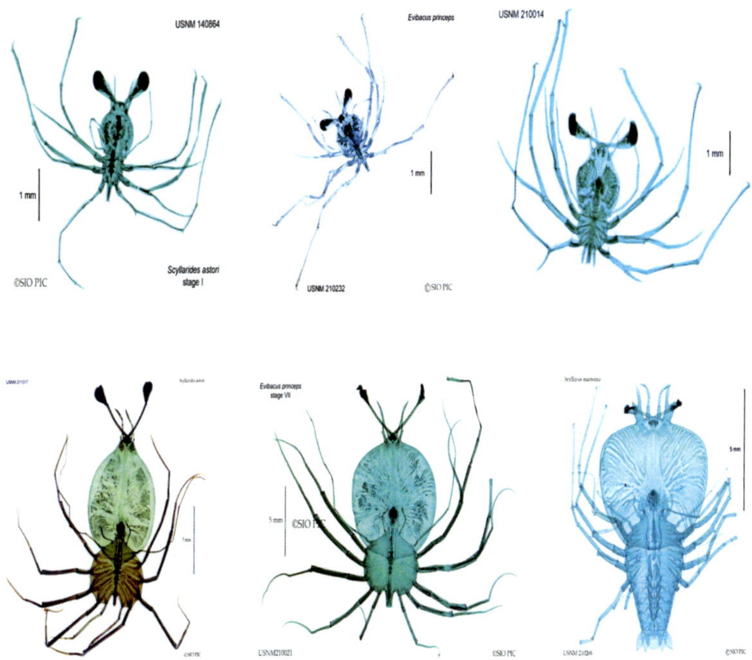

Figure 3.3. Synaxid lobsters.

Source: https://www.intechopen.com/media/chapter/68314/media/F3.png

The palinurid and synaxid lobsters, (Figure 3.3) scyllarid lobsters have garnered an interesting array of common names. These names, which come in many different languages, include slipper lobsters, shovel-nosed lobsters, squat lobsters, butterfly lobsters, locust lobsters, bugs and more. This variety is not due to confusion with other species of animals, but rather because the distinctive platelike antennae of slipper lobsters, which are shared by all 85 species of Scyllaridae named to date. This makes impossible to confuse them with any other decapod or, for that matter, any other crustacean. Despite having many more species than the Palinuridae and far outnumbering nephropid species, despite their distinctiveness being recognized throughout recorded history, they have not drawn the same level of scientific investigation as palinurid or nephropid lobsters. The lack of interest is presumably caused by their propensity for blending in with their surroundings, few of them are of substantial commercial significance, and the fact that it is convenient to classify them as Scyllaridae due to their homogeneous morphology and clear affinities. The Palinuridae and Synaxidae are undoubtedly related to the Scyllarid lobsters, despite their unusual body structure. Their long-lived, teleplanic phyllosoma larvae, a developmental form specific to these three families, are one of several notable morphological traits they have in common. The position of these three families within the Reptantia has been a major point of discussion in the past few years about the relationships between groups within the reptantian decapods. Despite this, the Scyllaridae's evolution has received little attention and efforts to further the taxonomic hierarchy beyond the family level have been made.

3.3 CLASSIFICATION

Following the separation of the Scyllarinae into 14 genera, Scyllarides has grown to be the family's largest genus with 14 species, followed by Scyllarus with 10 species (Cruz et al., 2021). A similar analysis of Scyllarus might lead to modifications and, possibly, the addition of new taxa to the subfamily. Although Arctides and Scyllarides, the two genera that make up the Arctidinae, are quite similar, they can be distinguished by their upper-surface structures. Arctides species exhibit a postorbital spine, a transverse groove on the first abdominal segment, tubercles and arborescent grooves on the abdominal somites. Scyllarides lack a postorbital spine and a transverse groove on the first abdominal somite; instead, their upper surfaces are smoothed by flattened, close-set tubercles that tend to cover the abdominal carina and conceal adjacent tubercles. The Ibacinae, which consists of three flattened genera, may be identified more quickly. With an armature resembling that of Scyllarides and a smoothed-over, squamose-tuberculate carapace and abdomen, Parribacus shows no signs of postrostral or branchial carinae. Evibacus has postrostral and branchial carinae that are equipped with rows of blunt teeth and is similarly squamose-tuberculate, albeit more delicately than Parribacus. On the carapace, it also features a characteristic posteromedian tubercle. The deep cervical groove is blocked, the eyes become uniquely encased in the Evibacus, and the third maxilliped's merus is enlarged and swollen with deep transverse incisions on its exposed surface. The carapace and abdomen of *Ibacus* species are smooth and, in some species, pubescent fine hairs; postrostral and branchial carinae are clearly visible, despite the absence of tubercles; and the eyes are located closer to the midline than the lateral borders. *Ibacus* species are the most flattened of all scyllarids.

Although there would seem to be a significant distinction between the two ibacinid families, scyllarid taxonomy has not yet formally recognized it. The anterolateral margin of the carapace is where the orbits of the monotypic subfamily Theninae are located and there are no projections of this margin below the eye. Additionally, the fifth pereiopod in females is achelate, the carapace lacks lateral spines or tubercles posterior to the post cervical spine, and it only bears a cervical incision on the lateral margin. The carapace of the Scyllarinae is more cylindrical than strongly flattened, has a cervical and post cervical incision, numerous teeth or tubercles on the lateral margin of the carapace, and the female has a chelated fifth pereiopod. The orbits are also not at the extreme anterolateral margin of the carapace, and this margin projects beneath the eye. Scammarctus, Bathyarctus, Antarctus and Petrarctus were divided into a group that lacked an arborescent pattern of smaller grooves and the remaining ten genera had arborescent grooves posterior to the transverse groove. The shape, profile and armature of the front margin of the thoracic sternum, as well as the presence or absence of a second carina or row of tubercles lateral to the single carina observed in all scyllarinid genera, on the fourth peduncular segment of the antenna, are used to make further divisions. The absence of the median dorsal carina on the abdominal somites in some genera is an intriguing development given that the presence of this carina is one of the family's defining traits.

There are several ways to examine evolution in the Scyllaridae, including the evolution of the Achelata within the Reptantia, the evolution of the family Scyllaridae within the Achelata, and the evolution of species groups as well as the species within the family. Data used to evaluate evolution include distribution, habitat, behavior and other biological features along with adult and developmental morphology, anatomy of living species, functional morphology of genes and other molecules, fossils and sperm morphology. With the help of this data, phylogenies are created that indicate potential origins, relative ages of taxa and distances between them. While many different traits are employed to categorize and classify taxa, their evolutionary importance varies, and their application, particularly with regard to morphological structures, is essentially arbitrary. Despite two genera, *Palibacus* and *Scyllarella*, being extinct, surviving species are among the oldest known species. However, some fossils are incomplete and difficult to correlate properly with extant forms. Because the carapace of *P. praecursor* is proportionately wider than of *Ibacus alticrenatus*, the mid-Cretaceous period likely saw the presence of an *Ibacus* species with a highly dorsoventrally flattened shape. The lateral expansions of the carapace, the anterolateral carapace angle, and the cervical groove of this extinct species, however, were distinctive in other ways. The extinct *Parribacus cristatus* species, found in northern Italy during the lower Eocene, resembles modern *Parribacus* species much more. The apparent absence of spines bordering the sixth and fourth antennal segments and the carapace, however, indicates that this specimen differs from existing species. The contour of these borders is broken only by a significant cervical groove. However, numerous other characteristics, including the shape of the borders, are noticeably comparable to those of *Parribacus* species that are still alive.

3.4 DISTRIBUTION

The distributions of the bigger scyllarids found in the families Arctidinae, Ibacinae, and Theninae are well known. The Scyllarinae contains a significant number of tiny species, some of which are poorly known and distribution data are frequently lacking. Since, it is the result of considerable thought and, with a few exceptions, seems to be broadly recognized, researchers believe the current taxonomy of the Scyllaridae to be trustworthy. The Scyllarinae were recently divided into genera based on more minor feature distinctions than those that set them apart from non-scyllarinid genera. This is where the most room for debate exists. Although classification does not always represent affinity, the Scyllaridae classification contains traits that are obviously important in evolutionary relationships. Nevertheless, it is likely that some convergence of forms is reflected in the current distributions. *Thenus* is once again the most obvious example, where the dorsoventrally flattened form is clearly adaptive to its sedimentary habitat, just as it is for *Ibacus* species. Despite having differences in maxilliped morphology that quite clearly place them in different lineages: *Ibacus* with the remaining Ibacinae and Arctidinae and *Thenus* with the Scyllarinae.

However, there is little paleontological evidence to support the notion that the Scyllaridae has experienced such isolations during its evolution or that current populations are remnants of former, more widespread taxa. Geological and habitat change always has the potential to isolate and divide populations. For instance, there is little evidence to conclude that the extinct

Scyllarella and Palibacus represent the dispersal of significant groups. Rather, it is more likely that the current distributions are the consequence of radiation and some specialization to take advantage of the available environments. While this seems to happen among scyllarids, different palinurid genera rarely inhabit geographical areas that correlate in latitude, depth or water temperature. However, compared to palinurid species, scyllarid species are more common and do not go as far into temperate latitudes. Instead, they are concentrated primarily in Australia, East Southeast Asia and, the Western Central Atlantic to a lesser extent. In addition, some palinurids are found at deeper depths than any scyllarid and may be better suited to colonize a wider range of environments than scyllarids. Additionally, the overlap of scyllarid genera points to the existence of mechanisms facilitating the exploitation of certain environments. *Thenus* inhabits sedimentary substrates, typically in waters less than 100 meters deep, whereas *T. orientalis* typically occurs at 10 to 50 meters depth. Such substrates are undoubtedly readily available given the genus's continuous distribution from eastern Africa to eastern Asia and Australia (Barth & Khan, 2008). Although *I. brucei* is also known from the northern central Tasman Sea and northeast of New Zealand, *Ibacus* species are typically found on shelf sediments at greater depths than *Thenus* and may be somewhat restricted to wide continental shelves adjacent to continents. In all of the world's major oceans, Arctides and Parribacus are coral reef and rocky shallow-dwellers. However, because they can be found in a wider range of depths and on a variety of surfaces, Scyllarides species cannot be as easily identified with habitat requirements. The genus does seem to have spread in the western Atlantic, but not significantly in the Indo-Pacific. This may be because of competition or because the environment in Southeast Asia is not ideal for their development. Although rarely found on muddy bottoms, it's possible that Scyllarides species are more frequently found in rocky terrain and coarser sediments.

3.5 GENETIC METHODS

The usage of molecular genetic methods has grown significantly during the past few years, along with public awareness of the importance of genetic information. The polymerase chain reaction (PCR), which enables the amplification of specific DNA sequences from little amounts of preliminary tissue, is a key component of many modern molecular genetic techniques. Due to its ability to process samples quickly, allows for non-destructive sampling and provide a wide range of specificity. PCR-based tests have quickly gained popularity and have replaced other methods as the primary strategy in molecular genetics research. The most popular method used in molecular phylogenetic investigations is the direct sequencing of PCR-amplified fragments. Sequencing provides the highest level of resolution, although it is a somewhat time and money-consuming phase. Genome size (GS) is a key characteristic of a particular species. Research on how it varies among living things is significant from both a theoretical and practical standpoint. The length of mitosis and meiosis in plants, body mass in mammals, annelids, flatworms, insects as well as copepods, duration of development in frogs, beetles, aphids and copepods are a few other traits that have been demonstrated to closely connect with GS. In addition, GS has been shown to be negatively linked with the rate of basal metabolism in mammals, passerine birds and amphibian brain morphology.

Figure 3.4. *Scyllarides latus.*

Source: https://previews.123rf.com/images/planctonvideo/planctonvideo1209/planctonvideo120900006/15488555-a-scyllarides-latus-a-pecies-of-slipper-lobster-shot-in-the-wild.jpg

Although not linearly, it was discovered that G+C concentration was significantly and positively associated with vertebrate genome size. Greater physical stability and some protection for genes against chemical damage are conferred by higher GC-percent values in larger genomes, which point to the existence of selection pressure for GC-enrichment of repetitive DNA. The ability of vertebrates to regulate their body temperature and the resistance of terrestrial as well as semi terrestrial mollusks to air exposure have been linked to DNA base composition. Studies on genome structure and evolution as well as genetic biodiversity require information on genome size and DNA base composition. DNA content has also been used to calculate weighing factors for ecological assessment, a quality metric for ecosystems. In addition, understanding the DNA content may be crucial for selecting model species for genome mapping and sequencing, or when organizing investigations involving hybridization and chromosome set alteration.

Figure 3.5. *Scyllarus pygmaeus.*

Source: https://www.atlantisgozo.com/wp-content/uploads/2015/10/Scyllarus-pygmaeus-Pygmy-Locust-Lobster.jpg

Only three Scyllaridae species, *Scyllarides latus* (Figure 3.4), *Scyllarus arctus*, and *Scyllarus pygmaeus* (Figure 3.5), have been cytogenetically examined. Giemsa staining, C- and fluorochrome bandings as well as other karyological techniques can only provide genetic information on the chromosome complement in terms of chromosomal size, shape and number due to a lack of specimen availability and technological limitations. In these species, gonads and hepatopancreas were used as the primary sources of chromosomal preparations. Decapod cell culture procedures have not generally been successful. Till date, only the *in vitro* maintenance of lymphoid organ cells of *Penaeus japonicus* was acquired but proliferation of these cells did not occur. Therefore, the only methods available for getting chromosomal preparations from crustaceans are direct procedures. The air-drying approach is specifically based on tissue dissociation, hypotonization and fixation of cells in suspension. Although several tissues can be used, the gonads, hepatopancreas, antennal gland and gills are the most appropriate in terms of mitotic activity. The testis is the most frequently employed tissue since it may produce a significant number of metaphases, but its activity depends on the season and stage of maturation.

Figure 3.6. *P. japonicas.*

Source: https://image.shutterstock.com/image-photo/fresh-kuruma-prawn-penaeus-japonicus-260nw-588473021.jpg

The extremely low number of metaphases and the presence of cellular debris in non-cultured cells are a drawback. Regeneration blastema in *P. japonicus* (Figure 3.6) was used to produce a sufficient amount of mitosis.

For small fragments of tissue or embryos, an alternate air-drying technique has been applied. Giemsa staining, which uniformly stains chromatin and allows for the study of the size, shape, the number of mitotic and meiotic chromosomes and heterochromatic chromosome bandings, allows identification of specific chromosomal regions, have both been used to obtain cytogenetic data on the Scyllaridae. C-banding and fluorochrome bandings have been used to study the heterochromatic areas, which are made up of DNA that is moderately and highly repetitive. C-banding is a method for locating and identifying heterochromatin in chromosomes. These regions may serve as highly helpful cytogenetic markers. Fluorochromes with AT or GC specificities can recognize heterochromatic areas with various base compositions.

Scyllaridae cytogenetics is still in its infancy. The limited specimens included in the reported data make it insufficient to describe the karyotype of the species under study and the few species under study prevent the inference of a broad pattern in chromosome evolution within the family. Overall, some observations can be made by contrasting these preliminary data with those of the Palinuridae species that are more well-known. The number of Palinuridae species that are diploid ranges from 111 to 158, with some species having the greatest 2n value. The number of diploid Achelata species ranges from 70 to 158 (Kampouris et al., 2021). The diploid number of *S. latus*, the only studied species of the subfamily Arctidinae, is 126 to 138, which is roughly twice the diploid number in the subfamily Scyllarinae. The diploid number of the Scyllaridae subfamily- Scyllarinae ranges from 70 to 74. A wide range of GS and a nearly consistent DNA base composition appear to be characteristics of the Scyllaridae family. Scyllaridae is a monophyletic group, therefore GS modifications during the evolution of this family appear to have happened without any preferential enrichment in GC or AT-DNA. GS values are currently more than three times higher in the subfamily Arctidinae than in the subfamily Scyllarinae. Because body size in the two Scyllarides species is larger than it is in the two Scyllarus species and ranks equally with GS among the examined species, the Scyllaridae appears to exhibit a positive association between GS and body size. Similar to the previous example, the biological variable Arctidinae appears to be positively connected to GS because it exhibits a longer larval life history than Scyllarinae.

Palinuridae is the other family in the infraorder Achelata for which information on GS and DNA base composition is known. Palinuridae C-values are intermediate within a narrow range than Scyllaridae, although the range of AT or GC-percent is greater. As a result, the two groups seem to have evolved their genome sizes along distinct paths. While, in contrast to the Scyllaridae, body size and GS appear to be unrelated, the existing data on life history within the Palinuridae further corroborate the proposed association between GS and length of larval life. For 84 species of Decapoda, the nuclear DNA content has been assessed. In this order, GS values are widely dispersed and slanted toward lower values, with the highest values typically relating to the infraorder Caridea. They vary from 1.07 to 38.00 pg with average 5.10 and are highly dispersed. The suborder Pleocyemata exhibits a similar distribution of GS values (Matzen da Silva et al., 2011). With the exception of Caridea, the infraorder Achelata of Pleocyemata has a GS range that is as wide as any other infraorder. The GS values in the genera Scyllarus and Scyllarides are among the lowest and highest, respectively. The AT-percent spans from 34.9 to 47.5 percent in the 12 decapod species that have been previously investigated, whereas all slipper lobsters share about 42 percent of AT-DNA, showing intermediate values.

3.6 DNA ANALYSES

Relative to a typical DNA content in cells from a reference species, the DNA content of target cells is measured. Samples are typically prepared using an internal reference in flow cytometric assessments of DNA content. Cell nuclei must be stained with a solution containing a certain fluorescent dye, buffering salts and a tiny amount of detergent, depending on the type of DNA content that will be analyzed. Although there are other available dyes that bind stoichiometrically to DNA, propidium iodide is the most frequently employed dye to quantitatively measure the size of the entire genome. For GC-DNA analysis, olivomycin or chromomycin A3 are suitable, while Hoechst 33342, Hoechst 33258, or DAPI are ideal for AT-DNA evaluation. The type of flow cytometer being used and its light source are crucial considerations in the stain selection process. Commercial argon laser flow cytometers typically do not provide the UV light that AT and GC-specific fluorochromes need to be excited at a specific wavelength. With the aid of particular optical filter sets, arc-lamp flow cytometers are more adaptable and can excite cells with UV or visible light.

Since few adult scyllarids are commercially significant due to their small size, there has been comparatively little research on their early life cycles. Some slipper lobsters are naupliosoma larvae upon hatching and all in the Scyllaridae family, along with spiny and coral lobsters, have a special, long-lived planktonic larva called a phyllosoma that develops into a nektonic/benthic post larva called a nisto. In comparison to taxa with small eggs, those with large eggs have stage I phyllosomas that are more developed. Some scyllarids, however, have developed limited larval lifetimes, so that with the emergence of specific behavioral patterns, larvae are not transported very far from the coast. There seem to be a lot of larval recruitment tactics between these two extremes. A wide variety of primarily pelagic fishes consume scyllarid phyllosomas, which appear to subsist primarily on soft, mushy meals. *Thenus* species members are extremely poor swimmers, with swimming abilities appearing to differ greatly across genera. Gelatinous zooplankton, especially medusae, is observed in association with several taxa that are in their late stages of development.

3.7 SLIPPER LOBSTERS

Slipper lobsters are closely related to coral or hairy lobsters and spiny or rock lobsters. Few scyllarid species are important commercially compared to several palinurid species, as *Jasus* and *Panulirus* species, which form significant, fully exploited fisheries, because most of their adults are small. As a result, fewer thorough studies of the biology and ecology of scyllarids have been conducted than those of palinurids, particularly on their early life histories. The synaxids, which have little to no significance for fisheries, are poorly understood. Scyllarids are common in temperate and tropical seas, on deep-sea ridges and seamounts, as well as tropical lagoons, continental shelves and slopes. In terms of size and habitat, they are a fairly diverse group. Many scyllarids have a highly flattened shape that is better suited to uniform, frequently soft substrates than the palinurids, which have a more vaulted shape. However, some scyllarids also have a vaulted shape and share the same habitats as the palinurids. The diversity of scyllarids shows that several larval recruitment techniques might exist. Although typically inadequate, primarily descriptive and frequently speculative, our understanding of the biology and behavior of the early life-history stages in the Scyllaridae does imply that there are many larval recruitment techniques. However, there are numerous early life-history traits which are shared by all scyllarid species that are now known. There are many similarities in the behavior and development of the larvae and post-larvae between the scyllarids and the palinurids.

The Scyllaridae undergo a sequence of free-living instars separated by molts during postembryonic development. The series of instars between these major molts, which include naupliosoma, phyllosoma, nisto, juvenile lobster and adult lobster phases, are referred to as phases in development. Phyllosomal instars may experience major morphological changes such as the development of appendix interna on the pleopods, which can occasionally be accompanied by a noticeable shift in behavior. Due to these morphological alterations, a new developmental stage has been identified. This stage can be distinguished by the degree of segmentation and appendage development. Each stage is typically connected to a single instar in species with few instars, such as *Thenus orientalis* i.e., each molt is accompanied by morphological changes interpreted as a change in the stage. In contrast, there will likely be more than one instar in the latter stages of species with long larval lifetimes, such as Scyllarides species. Therefore, only the stages of those species with long larval lifetimes can be determined with certainty for field-caught larvae, but the stage and instar number are clear for larvae raised in the laboratory.

Scyllarids can hatch as naupliosoma in some cases, but they all go through a succession of planktonic phyllosomal instars to mature. Large stalked eyes, uniramous antennules, biramous antennae, well-developed second and third maxillipeds and moderately well-developed first through third pairs of thoracic appendages are all features of the naupliosoma. Dorsoventrally flattened and transparent phyllosomas are suitable for planktonic life. Based on the species and stage of development, they have swimming exopods on some or all of their thoracic appendages, with pleopods being missing or only rudimentary until later stages. The phyllosoma that is in its terminal stage transforms into the nisto. The nisto, like the puerulus stage of palinurids, connects the planktonic larval and benthic juvenile phases of the life cycle by swimming to the settlement regions. The biology of phyllosoma, which is crucial for understanding larval recruitment, is poorly understood in many areas. Food and feeding, behavior, predation and sensory systems are some of these. Tank studies reveal that the phyllosomas of the majority, if not all, scyllarids are predominantly predators and use their pereiopods to fasten and hold food pieces. In nature, locations near water-land borders and coastal seas have higher densities of similar food items like larvae, jellyfish and small zooplankton than offshore, more oligotrophic waters. Many different fish species eat scyllarid phyllosomas, at least in their late stages. Scyllarid phyllosomas that remain close to shore during their development are particularly vulnerable to predation because many of these pelagic species are most frequently found in areas of high production, including coastal waters (Wakabayashi et al., 2014). As with the puerulus for palinurids, recruitment is finished at the nisto settlement, which connects the planktonic and benthic life phases in scyllarids. The nisto of at least some scyllarids, however, might be a more transitory stage and might resemble the first juvenile instar more than the puerulus. The universal observation that all of nistos are initially almost transparent, is a trait that has undoubtedly evolved to help them avoid predation as they move inshore. In contrast to *Thenus orientalis*, whose nisto completely lost its pigmentation at night but regained it during the day, the nisto of *Eduarctus martensii* remained transparent throughout the duration of the phase. *Petrarctus demani*, in contrast, displayed a slight loss of pigmentation at night. The latter two species' pigmentation changes probably help the nisto maintain its cryptic coloring against the shifting backgrounds that come with swimming in the water column and then resting on the substrate. Nistos are mostly active at night. Backward swimming, which is short bursts of oblique swimming controlled by bending the abdomen and telson, was a common night time activity of some scyllarids, such as *Eduarctus martensii* (Figure 3.7), *Petrarctus demani* and *Thenus orientalis* (Figure 3.8).

Figure 3.7. *Eduarctus martensii.*

Source: https://www.meerwasser-lexikon.de/imgHaupt/64556_5f8c7e991ce7a.jpg

Figure 3.8. *Thenus orientalis.*

Source: https://upload.wikimedia.org/wikipedia/commons/thumb/6/61/Thenus_orientalis.jpg/800px-Thenus_orientalis.jpg?20160329023521

The rapid beating of the setose pleopods allowed *P. demani* to swim backward as well. Some scyllarid species, including *Scyllarus americanus* and *Galearctus kitanoviriosus* have also been shown to forward swim by rapid pleopod thumping, which is often associated with time spent in the water column. The distinct variations in pleopod size and development among various scyllarids are likely related to differences in forward-swimming ability. For instance, the nisto of *Thenus orientalis* is probably not well-suited for forward swimming and directed shoreward movement because its pleopods are not sufficiently developed. In such nistos, the weak swimming ability suggests that the phyllosomas do not venture too far from adult grounds. The nistos of various other scyllarinid species, on the other hand, have well-formed, setose pleopods that each contain a conspicuous appendix interna that connects pleopod pairs.

Sand-burying during the daytime, a habit that palinurid pueruli are increasingly showing, may also be rather common in the nisto stage. As they travel toward their settlement grounds, nistos probably alternate between being active at night and hiding in or on the substrate during the day. Therefore, it is likely that the nisto is neither wholly nektonic nor entirely benthic. The nisto resembles the puerulus, wherein, it seems to be a nonfeeding stage that relies on energy reserves built up in the last phyllosoma stage. The animal may focus its efforts on selecting an appropriate settlement location since it no longer needs to eat. The Scyllaridae and Palinuridae families both include a special larval stage known as the phyllosomas (Woodings et al., 2019). Planktonic larvae with long appendages and flat, translucent bodies are known as phyllosomas. There is limited knowledge regarding how phyllosoma biology interacts with oceanic currents during this normally extensive dispersal period. Another significant environmental component influencing phyllosomal growth is salinity. The typical oceanic salinity level (35 ppt) is believed to be the ideal salinity level for growth because phyllosomas are commonly found in offshore waters. Despite phyllosomas' relative susceptibility to salinity fluctuations, suboptimal salinities, particularly lower salinity levels, result in delayed phyllosomal growth. Increased osmoregulatory needs and decreased food intake are suggested to be the causes of decreased growth at lower salinity levels. In general, marine crustacean larvae retain their hemolymph hyperosmotic to seawater via a mild hyperregulation. This is thought to be a process that provides the osmotic gradient necessary to achieve water endosmosis. The systems of hyper-regulation may need more energy at lower salinity levels to keep bodily fluids within phyllosomas at the proper osmolarity, which could delay growth. A longer intermolt time and a smaller rate of molt increase have been demonstrated to arise from reduced salinity levels in spanner crab zoeas, *Ranina ranina*.

3.8 PHYLLOSOMA

Many scyllarid species have long-living planktonic components called phyllosomas. For instance, the total body length of *Scyllarus* spp. changes from just over 1 mm to more than 30 mm during the phyllosomal period, which lasts for several months and involves numerous molting stages (Coutures, 2000). The preservation of these delicate, long-lived phyllosomas requires the design of an ideal tank that takes hydrodynamics into account for water circulation. Strong aeration can harm bodily parts because of their particular form. In the tank, there should be no water movement or very little water movement. Water flow must also take into account the phyllosomas' swimming, feeding and phototaxis behaviors, as well as the distribution of food and the accessibility of larvae in the tank system. An upwelling column of water that allows larvae to migrate vertically is referred to as a vertical water current. Although upwelling systems may require less floor space, the accessibility to the larvae is significantly limited. The horizontal water movement category includes the raceway and donut-shaped tank, in which larvae move horizontally while being driven by tiny water inlets. This is the only system capable of producing huge numbers of juveniles, despite the fact that the need for floor space is substantially increased. Accessibility to larvae is greatly enhanced, and there is space available to nurture many larvae.

Another important concern for the controlled environment raising of long-lived marine larvae is microbial control. The phyllosoma's natural environment is the open ocean, where the water is devoid of pathogens and low in nutrients. As a result, phyllosomas are extremely vulnerable to microorganisms in the water column. Historically, bacterial colonies in the raising water have been extensively controlled using a variety of antibiotics like oxytetracycline, streptomycin, chloramphenicol, etc. as well as chemicals. Due to microbial resistance, commercial-scale rearing of phyllosomas is not thought to be a viable long-term option, despite the fact that antibiotics and chemicals are powerful agents for reducing bacterial growth in the rearing water. Modern raising practices have adopted alternative sterilization techniques that significantly increase survival rates, such as the use of ozone (O_3) and ultraviolet light (UV) sterilizers. By directly irradiating microorganisms, UV light can both prevent microbial contamination of incoming water and maintain low microbial levels in recirculated rearing water. By oxidizing organic substances, ozone can also stop microbes from growing in recirculating water. Ozone should, however, be entirely removed before entering the raising water or must be maintained at low levels (0.05 ppm) because it is hazardous to all organisms, including phyllosomas. During long-term phyllosoma production, microalgae may also regulate the growth of bacteria and nutrients in the rearing water. The potential advantages of microalgae are still completely unknown, and there is very little information available on employing them to inhibit bacterial development.

For phyllosomas to be successfully raised, nutrition is crucial. The findings show that molting was delayed up to a point of no return for a period equal to the first starvation. The initial starvation has no lasting impacts on the next instar of phyllosomas in terms of larval survival, intermolt time or growth if they are fed during the PNR phase. The effects of subsequent starvation periods are visible on larval growth, but there are no corresponding effects on the next instar phyllosomas in terms of intermolt period and survival if feeding occurs after the point-of-reserve-saturation (PRS50) level, which is the initial feeding period at which 50% of larvae molted to the next instar(Zhang, Li, Li, & Xiang, 2014). These findings suggest that *Thenus* phyllosomas consume an energy reserve for molting before using any additional reserves for growth. The release of hormones created by specific endocrine systems regulates the molting process in crustaceans. Traditional androgenic glands and neurosecretory cells, which are specialized neurons that are a component of the central nervous system, are two types of endocrine organs that are connected to the molting process (CNS). The primary endocrine glands in crustaceans are the mandibular organs, which secrete methyl farnesoate (MF), crucial for metamorphosis and thought to be involved in the control of the Y-organs. The Y-organs, which secrete ecdysteroids like ecdysone that trigger molting. MF seems to be a hormone that keeps the phyllosoma in its immature state and prevents its metamorphosis, whereas ecdysone initiates the cascade of physiological and behavioral events that precede the molt. High levels of MF prevent larvae from growing and developing normally, and they result in intermediates that have characteristics of both larval and postlarval anatomy and behavior.

The sinus gland/X-organ complex, which contains the neurosecretory cells in crustaceans, is situated in the eyestalk (Rajendiran & Vasudevan, 2016). A number of neuroendocrine hormones important in molt regulation, such as molt-inhibiting hormone (MIH), mandibular organ-inhibiting hormone (MOIH), and crustacean hyperglycemic hormone (CHH), are stored and secreted by the sinus gland. The neurosecretory neurons that make up the X-organ are in charge of producing these hormones. MIH is the main hormone thought to prevent crustaceans from molting by inhibiting the Y-organs' ability to produce ecdysteroids. Fluorescence microscopy has been used to detect the sinus gland/X-organ complex in the phyllosoma of *Panulirus japonicas* as early as the first stage. In a different investigation, the expression of the MIH gene in sand crab larvae was discovered. These findings suggest that even at the earliest stages of larval development, at least some of these hormones are still active.

CHAPTER 4

Spiny Lobster Aquaculture

CONTENTS

4.1 INTRODUCTION

Shrimp trawls are used to catch lobsters and *Thenus orientalis,* which make up the majority of the bycatch,account for more than 70% of the lobster fishery(Meyer et al., 1998). Although they make up a minor portion of the catch in trawls, juveniles are caught in great quantities in nearshore areas using native fishing equipment such as fixed bag nets. Traditional gill nets and trammel nets, which are three-layered gill nets used to catch shrimp, are mostly utilized to catch lobsters around the southeast and southwest coasts at night. While *P. homarus* is the primary species landed in the southeast, followed by *P. ornatus*, the spiny lobster (*P. polyphagus*) and the slipper lobster (*T. orientalis*) make up practically the whole catch of the trawl fishery there. *Thenus orientalis* is the primary species caught in the southern trawl fishery. Prior to the development of the whole cooked lobster export trade to Japan in 1984, few fishermen fished for lobsters along the southeast and southwest coasts (Spanier et al., 2010). This led to an important increase in the value of small lobsters, which attracted more fishermen into the trade. The introduction of live export in the 1990s led to a rise in demand for little lobsters. There were more juveniles caught since many coastal villages participated in the fishery. India's government banned the export of spiny lobsters under 200 g and slipper lobster (*T. orientalis*) under 150 g in 2003 in an effort to reverse this trend. Juveniles still make up more than a third of the fishery despite this ban, which raises questions about the sustainability of the industry. In order to develop juvenile lobsters to market size in indoor tanks for value addition, a program for lobster culture was started in India.

4.2 DISTRIBUTION

P. homarus, *P. ornatus*, *P. polyphagus* and *P. versicolor*, the four principal species of spiny lobsters, were raised separately or collectively at densities ranging from 7 to 22 individuals/m^2. Spiny lobsters grew at a pace of 274 to 294 g per year during experiments carried out in indoor tanks at the Central Marine Fisheries Research Institute (CMFRI, Chennai) and the National Institute of Ocean Technology (NIOT, Chennai), with a survival rate of 90% (Kommuri et al., 2021). Despite this, commercial lobster farming was not started since it was unclear if it would be possible to stock the farms with enough baby lobsters to make a profit. There is an increase in the use of wild juvenile fish as well as shellfish for aquaculture and stock enhancement. Due to the lack of hatchery seeds, wild juveniles of fish species like eels (*Anguilla* sp.) and bluefin tuna (*Thunnus thynnus*) are collected as finfish. Due to the high expenses of seed production in the hatchery, spat of bivalves like oyster (*Crassostrea madrasensis*) and mussel (*Perna indica*) are also taken from natural grounds. One developing idea is that by avoiding excessive mortality during the early phases of larval settlement, the capture and cultivation of wild juveniles increase the overall fisheries productivity of the target species. Western rock lobster, *P. cygnus*, pueruli and postpueruli, death rates are high during the first year following settlement and range from 80 to 98 percent. Therefore, when significant numbers of pueruli are removed in low settlement years, the effect of puerulus removal on subsequent catches may be negligible. If a few numbers of post larvae and early juveniles are collected and grown without disrupting the normal recruitment to the standing stock, the total survival of a species can be considerably boosted. Based on this finding, attempts were made in Australia and New Zealand to cultivate lobsters using *Jasus edwarsii* pueruli that had been captured and kept in sea cages. To make lobster farming an economically feasible enterprise, numerous studies are being conducted on the settlement and collecting of pueruli, cage design and deployment, environmental consequences of growth and survival, feed composition, and other topics. Although collecting pueruli and early juveniles of spiny lobsters have been substantial commercial activity in Vietnam since 1990, the country's current yearly production is above 3000 metric tonnes. A program to raise lobsters in marine cages was started by NIOT in 2003 as a means of securing the livelihood of coastal fishermen along India's southeast coast (Manambrakat et al., 2009).

Figure 4.1. Juvenile spiny lobster.

Source: https://image.shutterstock.com/image-photo/juvenile-painted-spiny-lobster-panulirus-260nw-1485344870.jpg

4.3 JUVENILES

Raising juveniles of spiny lobsters (Palinuridae) (Figure 4.1) captured in the wild to market size is a growing aquaculture sector in Asia, Oceania and Central America, but Vietnam stands out as the region with the largest annual industry at US$90 million (Jeffs & James, 2001). A successful life cycle closure may finally be able to overcome the persistent danger posed by the sustainability of the wild juvenile fish supply to industrial growth. The current reliance on fresh fishery by-catch as a source of sustenance for lobster culture, however, is another and possibly more pressing barrier. The need for more environmentally friendly and sustainable pellet feeds has come to light as a result of declining catch availability, rising expense, and downstream environmental effects brought on by this sort of feeding. Until recently, efforts to provide tasty and effective pelleted dry feeds for spiny lobster grow-out were not very successful.

In several areas of the Asia-Pacific region, the tropical spiny lobster, *Panulirus ornatus*, is farmed in floating sea cages positioned in shallow coastal waters. The effects of this aquaculture activity on the environment are poorly understood despite its fast proliferation. In order to learn more about benthic carbon deposition and the production of dissolved inorganic nitrogen (DIN) from spiny lobster aquaculture, the integration of computer modeling with laboratory measurements is needed. Various feed conversion ratios (FCRs) employing wild seafood or an artificial lobster diet were used in modeling scenarios with two different lobster aquaculture stocking densities. The findings demonstrated that compared to artificial lobster feed, feeding lobsters seafood resulted in a much larger benthic carbon loading and release of DIN. Therefore, the environmental effects of spiny lobster aquaculture would be significantly reduced if waste fish were never used. Overall, the impacts of spiny lobster farming were geographically concentrated, with the highest levels of carbon deposition and DIN occurring directly below the sea cages. Therefore, unless the lobsters are overstocked and fed with feed, it is unlikely that spiny lobster aquaculture in sea cages will have a negative impact on the ecosystem.

Many Asian nations, including Vietnam, the Philippines, Indonesia, Malaysia, India and Thailand, have a significant spiny lobster (Palinuridae) aquaculture industry. *Panulirus ornatus*, *Panulirus homarus*, *Panulirus stimpsoni*, *Panulirus polyphagus*, *Panulirus longipes*, *Panulirus penicillatus* and *Panulirus versicolor* are just a few of the numerous species of commercially significant spiny lobster used in cage aquaculture in sea. The most valued variety of lobsters among these is the ornate tropical lobster (*P. ornatus*). Wild-caught young lobsters are fattened to a marketable weight of between 800 and 1000 g while being kept in sea cages. Either prepared artificial diets or, more frequently, garbage fish are used to feed cultured lobsters. Vietnam currently produces 1900 ton of spiny lobsters annually using marine cage aquaculture, with a market value of more than US$65 million in 2006. The bottom and the water column may suffer as a result of the enrichment of nutrients brought about by the culture of marine species, such as finfish, in sea cages. While excretion from aquaculture animals causes dissolved inorganic nutrients to be released directly into the seawater, sources of nutrient enrichment from cage aquaculture in sea include fecal material from undigested feed and waste feed that will eventually sink to the seabed as particulate organic material. In extreme cases, excess organic carbon on the seabed can cause the formation of bacterial mats, the release of methane and hydrogen sulfide from the seabed, and the promotion of anoxic conditions in sediments, all of which can significantly reduce the diversity of the benthos. Dissolved inorganic nitrogen is one of the primary excretory products from protein metabolism by cultured species. In addition to encouraging the growth of microalgae, the increase in DIN in seawater can also be hazardous to marine life in high quantities.

4.4 AQUACULTURE AND ITS IMPACT

Despite the fact that *P. ornatus* cage aquaculture in sea has been spreading quickly over portions of Asia, less is known about the effects of this practice on the environment. Based on the type of meal being provided to the spiny lobsters, previous studies suggested that the generation of *P. ornatus* in sea cages resulted in the release of an estimated total of 269–402 g N/kg into the environment (Priyambodo et al., 2020). It is challenging to even start figuring out the fate and effects of these nutrients discharged into the environment in various ways since these data do not differentiate between organic nitrogen i.e., bound in fecal material and waste feed and DIN i.e., from expelled metabolic waste. Furthermore, while being a crucial metric for figuring out the expected environmental effects of this activity on sediments, the organic carbon deposition on the bottom from the *P. ornatus* sea cage operation has not been determined. In order to predict the likely environmental effects of spiny lobster sea cage operations using both artificial diet and seafood diet, as well as for a variety of stocking densities and FCR, researchers used hydrodynamic modeling in conjunction with earlier laboratory measurements of nutrient outputs. Environmental data from a field site suitable for establishing this activity was incorporated into the spiny lobster cage cultureoperations in sea that were modeled, along with plausible spiny lobster farming scenarios based on variations on the currently emerging commercial practice for this activity in some regions of Asia. One of the difficulties in contemporary aquaculture is the closed-cycle rearing of palinurid lobsters in commercially relevant quantities. The main barrier to their successful aquaculture is the length and complexity of the larval life cycle, which exacerbates the issue.

Figure 4.2. *Palinurus elephas.*

Source: https://upload.wikimedia.org/wikipedia/commons/thumb/1/1f/Monaco.Mus%C3%A9e_oc%C3%A9anographique089.jpg/800px-Monaco.Mus%C3%A9e_oc%C3%A9anographique089.jpg

It is widely believed that advancements in the crucial field of nutrition research will lead to the innovations required to make closed-cycle lobster raising a reality. There isn't now a commercially available feed for their culture because of the enigmatic nature of their eating preferences and complicated larval anatomy. However, studies have been done to clarify many of the unanswered questions about the nutritional needs of larvae.

Researchers infer that the quantitative nutritional needs of larval spiny lobsters are probably lower than those of other larval species. The accumulation of lipids to support an energy-demanding metamorphosis and a later nonfeeding puerulus phase ultimately seems to be what determines the overall success of the larval cycle. These realizations offer insightful guidance for the development of nutrient-complete diets to enable the closed-loop culture of spiny lobster species.

Palinurus elephas, (Figure 4.2) a cool temperate species from Europe, *Jasus lalandii*, *J. edwardsii* and *J. verreauxi*, cool temperate species from the southern hemisphere, and *Panulirus japonicus*, a warm temperate species from Japan, have complete larval development of the phyllosoma stages. Wild males and females were brought to Sanriku, Japan, where they were raised there for a number of years in a row. Every year, between spring and summer, there was one hatching. Water moved back and forth between the larval culture container and a 200 or 500 L microalgae *Nannochloropsis* spp. culture tank. Food for the early phyllosomas included bits of mussel (*Mytilus edulis*) gonad and Artemia nauplii. After the second or third instar, just the mussel gonad was used. With *J. verreauxi*, the larval survival rates were highest, but *P. elephas* had the lowest rates. Despite this, *P. elephas* is regarded as the most promising aquaculture species because of its short phyllosoma life of 132 days compared to the other species' 200-300 day lifespan (Kittaka et al., 2001). There is a lot of interest in the aquaculture of spiny lobster because of concerns about the sustainability of their fisheries. The availability of inappropriate foods throughout the protracted phyllosoma larval period has hindered attempts to culture spiny lobsters from eggs to puerulus, which has been generally ineffective. The size, texture and nutritional components of a suitably prepared diet will be determined by a better understanding of the feeding behavior, natural dietary preferences, nutritional profiles and digestive capacities of phyllosoma. The most promising designed diet for early-stage larvae appears to be soft, perhaps microencapsulated meals, followed by fleshy prey in later stages, according to current morphological and behavioral research reported.

Beginning in early 2000, some farmers began working at various locations in the southern part of the island of Lombok to grow little spiny lobsters. In the region of the subtidal zone, undersized wild-caught lobsters were put in floating cages and fed garbage fish. They were picked when they weighed 150–300 g and were marketable after 7-8 months. Despite relying on wild-caught seed, lobster cultivation continues. After some farmers were successful in growing these juvenile lobsters to a marketable size, the act of catching young lobsters became a profitable sector of the industry chain. *Panulirus homarus*, *P. ornatus*, *P. longipes*, *P. versicolor*, *P. polyphagus* and *P. penicillatus* are among the spiny lobster species exported from Indonesia. The aquaculture of lobster in floating net cages has significantly boosted the market in the southern region of Lombok, where the youngsters were discovered. Around 4-5 tons of lobsters are produced each month by more than 1,500 small-scale farm operations. The lobster aquaculture industry will continue to be successful due to the robust market for lobsters and the success of lobster growth. Compared to lobsters produced by lobster farmers, traders choose larger lobsters. There isn't much of an incentive for farmers to create larger (1 kg) lobsters because the current lobster farming techniques are making small-scale farmers significant profits.

4.5 SUPPLY

With high demand and constrained supply, the spiny lobsters *Panulirus ornatus* and *Panulirus homarus* are significant tropical aquaculture species under development. A formulated feed must be developed because the established sector in Vietnam relies on wild-caught mixed seafood bycatch as feed, a practice connected to water quality degradation and potentially disease proliferation in lobster farming. The lack of the crustacean and mollusks components of the seafood bycatch used in Vietnam increases the demand for synthetic feeds in the developing Indonesian and Australian sectors. Knowledge of nutrient requirements, component quality, physical feed requirements and the relationship between feeding behavior as well as feeding methods are all necessary for the formulation of such feeds. Instead of influencing commercial least-cost formulation techniques, research so far has concentrated on creating a feed recipe that will be ingested and maintain adequate growth. Future research will undoubtedly be necessary to inform formulation, but in order to ensure that such research is used, it is also important to comprehend the diverse expanding lobster farming sectors and the factors influencing their acceptance of formulated feeds. To make sure that the research can be put into practice and to address farmers' perception of formulated feeds, this will require collaboration throughout the supply chain. While validation through semi-commercial benchmarks and demonstration farm models are expected to improve commercial uptake of produced feeds, technical aspects of feed manufacture and scale-up of feed advances will be crucial for the adoption of research results. Due to their high demand and constrained worldwide supply, spiny lobsters are among the most expensive seafood in the world. As the primary source of sustenance, lobster farming mainly relies on wild-caught shellfish. This can be found in Vietnam and reflects the normal feeding choices of lobsters, which also include fish, crabs, mussels and echinoderms. While ensuring that meals are tasty and completely nutritional, this approach has significantly degraded water quality, which can render the environment less suited for lobster culture and cause disease. The lobster aquaculture sector has been highlighted as facing significant perceived constraints related to water quality as well as access to suitable and economical feed, both of which have prompted the industry to develop and use formulated diets. It is crucial to create formulated feeds in order to maintain sustainable development with the growth of the lobster industry in Vietnam and the burgeoning sector in Indonesia. Throughout their life cycle, spiny lobsters experience major changes to their body structure, feeding habits and digestive physiology. Puerulus is a stage of life in which Phyllosoma larvae do not feed; they are specialized for planktonic feeding. *Panulirus ornatus* and *Panulirus homarus* are the two species that make up the majority of the puerulus supply in Southeast Asia, with the former dominating recruitment in Vietnam and the latter being more prevalent in Indonesia. *P. ornatus* has been the subject of recent commercial hatchery development in Australia (Jones & Shanks, 2009).

4.6 FEED

Formulated feed development for lobsters is still in its infancy, with a lot of attention being paid to improving physical condition and appetite. A variety of mouthparts, chemo-sensing and manipulating appendages work together to decide what is consumed by crustaceans throughout the intricate process of digestion. The function of feeding appendages and mouthparts that result in ingesting must therefore be understood in order to maximize feed intake. Members of the genera *Panulirus* and *Jasus* share a lot of similarities in their mouthpart and foregut morphologies. The mouthparts of spiny lobsters, both juvenile and adult, are responsible for primary feed, feeding into the esophagus, and generating respiratory currents over the gills. The main appendages for selecting feed to be delivered to maxillipeds for initial feed size reduction and additional manipulation appear to be periopods. Larger particles may travel anteriorly to two sets of maxillae while smaller particles may be carried toward the mouth opening by water currents created by the three sets of maxillipeds. The maxillae have setae and spines that are utilized to select and drive food into the mandibles while anteriorly eliminating unwanted food particles. The molar and incisor processors in the mandibles, which are severely calcified, use shear pressures to reduce the size of big, calcified material when it is swallowed before further trituration in the gastric mill. External size-reduction of food components causes waste feed particles to escape the mouth or to be rejected without ingestion, which results in inefficient feed intake and utilization that can be decreased by using the proper size and shape formulations for the feed.

Anterior or cardiac and posterior or pyloric parts make up the chitin-lined foregut or proventriculus of post-puerulus lobsters. The gastric mill, which is made up of ossicles in the anterior region, is in charge of reducing feed particle size after ingestion. The cardiopyloric valve permits small particles to pass through and allows larger particles to interact with the enzymes from hepatopancreatic gland and be processed again through the gastric mill. A filter press ensures that only very small particles and liquids enter the midgut gland for additional digestion and absorption after leaving the posterior foregut. Feed particles must be able to be ground to a fine size before digestion in order to comply with this method. Due to the best growth of *Litopenaeus vannamei* when feed particle sizes average 124 m prior to feed manufacturing, fine grinding of feed raw materials has traditionally been employed for shrimp feeds. Indigestible particles larger than 100 nm are excluded from the midgut gland (hepatopancreas) of *Penaeus monodon*, showing that better growth associated with fine grinding is probably caused by more quickly triturating and digesting feed rather than facilitating the direct transfer of feed particles into the midgut gland.

However, it is still unknown if fine grinding is necessary for optimal feed intake and growth in spiny lobsters. It is interesting to note that lowering the particle size of fish meal to 500 m failed to increase feed apparent digestibility for *J. edwardsii*. Foregut capacity is thought to be what restricts lobsters' ability to consume formulated feed. This capacity is probably regulated by how much feed expands after being consumed and how quickly it exits the foregut. The capacity for ingesting new feed is decreased by the fact that extruded feeds frequently expand significantly after being hydrated and may consequently occupy more space than pressed pellets. Gut passage rate is probably correlated with raw material particle size given the size-selective recycling of feed particles through the gastric mill. Therefore, particle size and digestion rate are probably crucial for maximizing feed intake capacity and reviving hunger in between meals. Since the apparent digestibility coefficient (ADC) and gut transit rate can be inversely related, suggesting that slower rates of gut transit promote more thorough digestion, the rate of digestion is independent of the apparent digestibility coefficient (ADC) in this sense (Nankervis & Jones, n.d.). In contrast, water-soluble components like sodium caseinate and hydrolysates are believed to speed up the transport of feed to the gastric gland, making them quickly available for assimilation and growth and freeing up foregut space for additional feed ingestion. In an effort to provide a low carbohydrate feed with significant levels of soluble protein from sodium caseinate, this method was invented for *J. edwardsii*. However, it is included in recently published reference diets for the subtropical lobster *Sagmariasus verreauxi* and tropical slipper lobsters *Thenus australiensis*, with almost 50% of the feed consisting of sodium caseinate. The application of this concept seems to have been limited for this species due to the importance of carbohydrates as an energy source and relatively low protein requirement.

4.7 GROWTH

The growth of *P. ornatus* appears to be linearly correlated with the amount of digestible protein present in the meal, at least for feed containing up to 56% digestible protein (61 percent crude protein. This suggests that it may be incorrect to generalize nutritional criteria across spiny lobster species. In order to determine the proper dietary requirements, non-target growth-limiting variables must be eliminated. Future feed development for *P. ornatus* and *P. homarus* will depend on properly satisfying these high protein requirements with suitable raw materials. Although lobsters are a particularly desirable aquaculture product, a cost-benefit analysis will be required to understand the trade-off between growth and dietary protein levels in commercial feed production. This is because protein-dense components are expensive.

Because of the frequent use of high amounts of fish meal or other marine proteins and the belief that the marine protein component will satisfy all required amino acid requirements, amino acid requirements have not been a research focus for studies on lobster nutrition. Prawn meals provide the most accurate information for essential amino acid needs since requirement values there have been well investigated and are consistent across species. While commercial shrimp contain protein levels close to those needed by *P. ornatus*, kuruma prawns have a minimum protein need of 42 percent. However, appetite stimulation and growth rates seem to be greater when fresh mussels are the predominant feed material comparable with observations with *J. edwardsii*. These kuruma prawn diets appear to be among the highest performing feeds for *P. ornatus* (Nankervis & Jones, n.d.). A significant factor in the success of these proteins as food for *P. ornatus* is probably the high protein content of mussels and other reference marine products like squid and fish. Although mussels and other reference proteins appear to have lower minimum essential amino acid requirements than kuruma prawns, the amino acid pattern in relation to lysine level is consistent between the kuruma prawn requirement value and the reference proteins and is likely to be a good reference for the advancement of our knowledge on essential amino acid requirements. The ratio of essential to non-essential amino acids, as well as the total amount of each individual amino acid, may act as a mediating factor in the required values. When compared to fish meal, squid and mussel proteins stand out for having a particularly high percentage of arginine, which may explain why mussels are so successful as a sole source of protein for spiny lobsters.

Since it is assumed that the culture water and feed already contain enough minerals to meet needs, no research has been done to date on the mineral requirements of spiny lobsters. Crustaceans need calcium, copper, phosphorus, potassium, magnesium, selenium and zinc. However, it is unknown how much of these nutrients should be added to practical feeding. Since, periodic mineral supplementation may be necessary due to ecdysis mineral loss, meals should contain some amount of each of these minerals. In crustaceans, phosphorus and calcium perform crucial structural and functional roles, and their requirements are interconnected. While calcium is plentiful and can be taken from the water, phosphorus is often in little supply in high-quality seawater. Although calcium is not necessary for marine crustacean feed, it is frequently present in raw materials and can limit phosphorus availability, leading to recommendations for calcium: phosphorus ratios of less than 2:1 and, in certain situations, less than 0.5:1. With phosphorus requirements for crustaceans ranging from 1% to 2% of the feed, there is a wide range of potential minimum phosphorus requirements as well as maximum calcium requirements. Vitamins are necessary for healthy growth, proper metabolism, and survival. Although their needs have not yet been measured, lobsters are not an exception. All vitamin requirements for shrimp species have been measured and equivalent requirements for lobsters are hypothesized. For lobsters feeding, a diet mostly composed of wild fish, fish meal and wheat, vitamin supplementation is necessary for optimum development and FCR, suggesting that wild fish may not be sufficiently vitamin-rich for optimal output. While the needs for many vitamins vary little from one species of crustacean to another, those for vitamin C, choline and inositol are very different.

Figure 4.3. Penaeid shrimp.

Source: https://upload.wikimedia.org/wikipedia/commons/thumb/9/98/Penaeus_monodon.jpg/800px-Penaeus_monodon.jpg

As has been demonstrated for penaeid shrimp (Figure 4.3), astaxanthin accumulation in lobster tissues suggests a potential antioxidant and health function of this carotenoid, which may promote resilience to stresses. Although the advantages of higher supplementation levels are still unknown, astaxanthin is necessary at 50 ppm for ideal carapace color. According to additional studies, feed with 800 mg/kg of astaxanthin produced growth that was 55% higher than feed containing 100 mg/kg. According to anecdotal evidence, 100 mg/kg of carotenoids increases development and survival by lowering oxidative stress in the clawed lobster *Homarus gammarus* (Ambati et al., 2014). Future studies should clarify whether this effect is constant, whether it is unique to astaxanthin, or whether it may be remedied by antioxidants that are more reasonably priced.

Spiny lobsters' growth and health appear to be best served by manipulating their gut flora. The microbial community, growth rates, and gut health of *P. ornatus* and *P. homarus* can all be supported by mannan oligosaccharide (MOS) feeding. In response to MOS supplementation, *P. ornatus* gained more than 50% more weight, had enhanced survival rates and had a 10-fold increase in the quantity of aerobic bacteria in their diet of frozen fish and prawns. Both experiments used very small lobsters, with final weights of 2.9 and 4 g for *P. ornatus* and *P. homarus*, respectively, even though MOS consistently promoted growth and intestinal health for both species. Validation for larger lobsters may be worthwhile for efficacy and cost optimization since it appears that the optimal incorporation level of MOS decreases with lobster size. For *P. ornatus*, probiotics have been studied, with an emphasis more on bacterial disease resistance than growth effects. Probiotics seem to be successful at controlling Vibrio, but their use in feeds can be challenging because of their heat sensitivity, which restricts them to cold-pressed pellets or post-processing applications. Given the effectiveness of MOS, it is crucial to refine techniques for influencing gut flora for the best growth and health advantages. For instance, sodium propionate increases feed intake, growth and the digestibility of phosphorus and energy, while salts of organic acids have demonstrated effectiveness in suppressing Vibrio in shrimp feeds.

4.8 CHEMORECEPTORS

Decapods have chemoreceptors on their walking legs and antennae, which they utilize to distinguish between food and other particles. These sensors in homarid lobsters are especially sensitive to mussel extracts, but they are also activated in varying degrees by a variety of other small nitrogenous compounds, including L-glutamate, hydroxyproline, ammonium chloride, L-aspartate, L-arginine, glycine, glutathione, betaine, and taurine, in decreasing order of significance. To boost feed consumption, these nitrogenous substances can be added to lobster meals in the form of fresh seafood, krill meal, fish hydrolysates or refined versions of chemicals such as amino acids or betaine. The two main water-soluble substances from mussel meat that are most likely to help lobsters sense their food are taurine and glycine. Compared to pelleted feeds, which are only consumed over a period of 2-3 hours before the attractant properties of the pellets are lost or the foregut capacity of lobsters limits further ingestion, both compounds are hypothesized to contribute to the extended feeding response documented for mussels. Products made from krill are thought to be especially crucial for ensuring high feed intake and enabling the creation of a dry feed for subadult lobsters without the requirement for fresh seafood components. Smaller (1 g) lobsters, however, seem to react to fresh seafood products significantly more strongly. It is unclear if this reflects an ontogenetic change in feeding preferences, variations in the raw materials employed or differences in the experimental setup. It is obvious that more study is needed to determine the factors that influence hunger and feed intake in spiny lobsters. Protein hydrolysis, which solubilizes the protein and increases the number of binding sites between amino acids and lobster chemoreceptors, has been a frequent technique for making protein supplies more appealing. A variety of fish and crustacean hydrolysates have been found to boost feed intake over a two-week period, according to further research, but it is yet unknown whether this results in faster growth. To maximize growth rates and the financial viability of commercial lobster feed production, it will be critical to define criteria for hydrolysates or other hunger stimulants that consistently boost growth.

The development of feeds for lobster research to date has taken two totally different paths, but both have as their common goal the formation of a matrix that enables adequate feed intake. The more conventional approach to feed formulation began with a fish meal foundation and a starch binder. To improve pellet binding, a synthetic ethylene/vinyl acetate copolymer pelleting agent was used, and the result was a semi-moist pellet. Although mixed fresh seafood has been added for flavor, krill meal has lately been found to be more beneficial at promoting the growth and survival of *P. ornatus*. For slipper lobsters and the temperate spiny lobsters *Sagmariasus verreauxi* and *J. edwardsii*, an alternative strategy based on sodium caseinate has been proposed (Nankervis & Jones, n.d.). This is likely because sodium caseinate has a highly concentrated water-soluble protein and can be bound with transglutaminase to form protein cross-linkages rather than relying on starch for binding. The application of transglutaminase and sodium caseinate to lobster feeds allows for additional room for the high protein need. This combination is widely employed in food technology to recombine meat products and alter the textural qualities of protein-bound food.

4.9 MUSSEL MEAL

Green shell mussel (GSM) (Figure 4.4) meal and betaine provide the feed with protein content as well as palatability whereas, krill oil boosts lipid content by combining with lecithin to provide 1.39 percent of the feed as phospholipid. The importance of phospholipids has been studied, but it is very challenging to evaluate their significance given the experiment's incredibly slow growth rate; only 4.5 percent of body weight was gained over the course of 50–55 days (Miller et al., 2014).

Figure 4.4. Green shell mussel.

Source: https://niwa.co.nz/sites/niwa.co.nz/files/styles/sidebar/public/IMG_2009.jpg?itok=fvnlsGJE

One percent Carophyll Pink, 8 percent or 800 ppm astaxanthin and one percent lecithin, is present in both reference diets. They differ slightly in how much cholesterol they include, demonstrating the need for even further cholesterol optimization. High protein and moderate fat requirements have been established for both reference meals. Both of these feeds contain ingredients that are not typically found in commercial formulations or at least not at the same inclusion level. This is due to the cost of the ingredients, even though they are appropriate references for future development and incorporate various aspects of the knowledge amassed to date. Before their commercialization is feasible, the implementation and uptake of feed formulations for lobsters are expected to require significant cost optimization. While both of these reference diets currently support growth that is less than that of fresh mussels, their potential for use in regions where fresh seafood bycatch is not easily accessible seems promising. Both feeds serve as the end result of the research that has been done in the production of lobster grown out feed up to this point and serve as a foundation for future study.

Despite the fast growth of spiny lobster aquaculture industry, little is known about its possible environmental effects. The fecal output and settling velocity of spiny lobster feces were measured in an effort to comprehend the potential effects of sea-cage cultivation. In order to model various scenarios of farm stocking density and food conversion efficiency and draw conclusions regarding possible benthic repercussions, these data were then employed in hydrodynamic modeling to analyze the dispersal pattern of fecal material from two sizes of hypothetical spiny lobster farms. The average lobster fecal strand settling speed was 0.9±0.05 cm/s (Lee et al., 2014). The weight and density of fecal threads and their settling velocity were positively correlated. On the other hand, there was no correlation between the settling velocity and the length or surface area of fecal threads. Fecal material deposition was primarily found directly beneath the sea-cages, according to hydrodynamic modeling. Higher stocking densities and FCR values were correlated with an increase in the density of fecal material deposition. Overall, the findings show that, even when operating at the highest commercial stocking and feeding levels currently available, the deposition of fecal waste from sea-cage lobster aquaculture is unlikely to have a significant negative influence on benthic ecosystems.

In a number of Southeast Asian nations, the spiny lobster, *Panulirus ornatus*, is a highly prized seafood item. However, compared to other Palinurid lobster species, *P. ornatus* is rather uncommon in the wild, therefore developing an aquaculture industry for this species could provide a way to balance supply and demand. The development of closed-life cycle production technologies is crucial for the agricultural sector to be truly sustainable. The larval phase (phyllosoma) of *P. ornatus* has recently been successfully completed in captivity on multiple occasions, yielding enough data to create an exhaustive morphological description. A morphological comparison is done between wild specimens and hatchery-raised phyllosoma. The size of phyllosomas from captive-raised and wild-caught fish was similar. In contrast to estimates of wild phyllosomas, it was shown that some hatchery-raised individuals experience abbreviated development and have the capacity to undergo metamorphosis more quickly than expected. On the plasticity of the larval duration in *P. ornatus*, observations are conducted. The phyllosoma of *P. ornatus* from both captive and wild populations showed up to total of 24 morphological increments. By identifying the beginning and end of particular morphological characteristics, these were classified into 11 different stages. This descriptive morphological key offers a unique point of reference for tracking this species' larval development. The varied

P. ornatus larval duration shows that the length of the hatchery phase may be greatly reduced by optimizing husbandry and feeding conditions, increasing the likelihood of producing seedstock for an aquaculture sector based on closed-life cycle for spiny lobster culture.

These measurements also evaluated the ability of newly-hatched phyllosoma to digest food when switching from endogenous to exogenous nutrient sources. The sequential use of fat, protein, and carbohydrates by embryos was revealed by ontogenetic variation in digestive enzyme activity. Non-specific esterase activity appears to increase early in embryonic development, suggesting that lipid is the most important macronutrient, with protein and carbohydrates being used more and more before and after hatching. It is likely that the observed surge in trypsin-like protease activity on the first day after hatching enables phyllosoma to utilize foods high in protein at first feeding. Changes in digestive enzyme activity after hatching indicate that a metabolic switch from protein to lipid may have taken place, corresponding with the date of the point-of-no-return for Stage I phyllosoma of *P. ornatus*. They emphasize the value of giving phyllosoma an exogenous protein-rich food source immediately after hatching.

The hemolymph in the open circulatory system of lobsters contains bacteria. As the closed life-cycle aquaculture of the spiny lobster *Panulirus ornatus* gets closer to being a commercial reality, it becomes more crucial to understand the function of these microorganisms, especially for the diagnosis of disease. Lobster survival, total granulocyte counts, and the diversity, load, and functional potential of the hemolymph bacterial population are all impacted by the elevated temperature. Proteobacteria and Bacteriodetes make up the majority of the hemolymph microbiome, according to pyrosequencing analysis. Rhodobacteraceae members were heavily represented in both studies, although overall, culture-independent approaches caught a larger bacterial diversity and load when compared to culture-dependent methods. It has the potential to be employed in health monitoring systems as it is the first thorough study to compare the hemolymph bacterial populations of healthy and thermally stressed cultured juvenile *P. ornatus*.

CHAPTER 5

Diseases of Lobsters

CONTENTS

5.1 INTRODUCTION

Diseases that affect crustaceans' survival and ecology can have further negative economic effects on fisheries and aquaculture. There have been many parasites identified in crustaceans, and the finding of new parasites has become more effective with the development of cutting-edge technology like next-generation sequencing (Stewart, 1984). Although they frequently lack information on pathophysiology, molecular approaches are starting to outperform more traditional technologies for parasite discovery. Histopathology is still one of the most affordable techniques for finding parasites and enables both parasite identification and descriptions of the pathology they produce. The strategy is effective when combined with current molecular and electron microscopy methods, however, there aren't many resources available to interpret histological slides from crustaceans (Shields et al., 2012). Early works were restricted to lesser resolution than what is currently available, and those that are currently available do not offer comprehensive images of all organs. The histology reports include comprehensive descriptions of infection, but few also include photographs of healthy tissue or a description of a baseline for comparison. The histology of the animals exhibits high visual similarity across all tissue types.

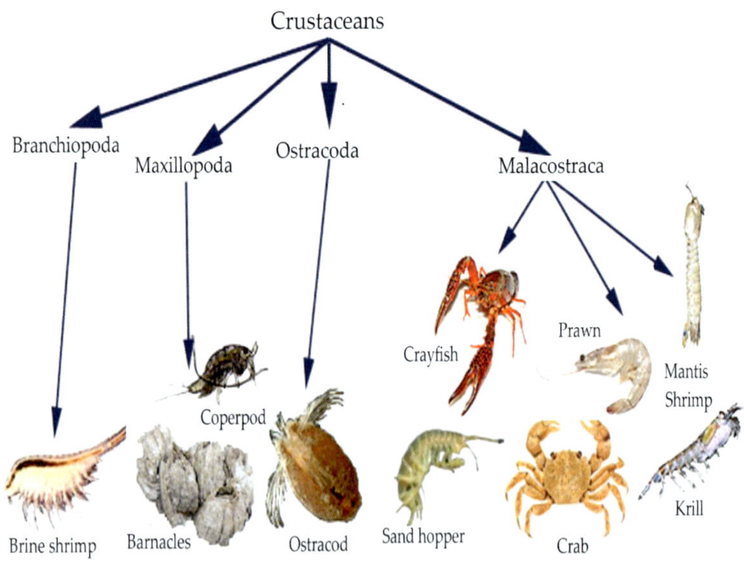

Figure 5.1. Crustaceans.

Source: http://www.mesa.edu.au/crustaceans/images/intro.jpg

5.2 CRUSTACEANS

Crustaceans (Figure 5.1) are a significant component of the world's fishery as well as aquaculture species and play significant ecological functions in both terrestrial and aquatic settings, contributing up to $40 billion USD in yearly retail value. The relevance of comprehending crustacean disease outbreaks in the fisheries and aquaculture industries is highlighted by the increased reliance on marine settings for food supplies. Decapod crustaceans, which are significant commercially, have seen some of the most famous mass extinctions of the past fifty years. The most notorious virus to affect penaeid aquaculture is the White Spot Syndrome Virus (WSSV), (Figure 5.2) which has destroyed shrimp aquaculture globally and been found in a number of different decapod hosts.

Figure 5.2. White Spot Syndrome Virus.

Source: https://www.outbreak.gov.au/sites/default/files/images/prawn-white-spot.jpg

Figure 5.3. Portunid crab.

Source: https://upload.wikimedia.org/wikipedia/commons/thumb/2/2f/Liocarcinus_depurator.jpg/800px-Liocarcinus_depurator.jpg

The portunid crab, (Figure 5.3) crabs in the genus *Chionoecetes* are all species of commercially significant crab and lobster that are known to be infected by *Hematodinium*, a genus of parasitic dinoflagellate.

Furthermore, epizootic shell disease (ESD) (Figure 5.4) in American lobsters (*Homarus americanus*) has both sublethal and lethal consequences that can adversely affect larval recruitment, with ovigerous females having the highest prevalence of the disease. Such diseases deplete the energy reserves of crustaceans and frequently cause host mortality, decreased fecundity or decreased growth. Despite the fact that the majority of crustacean diseases do not directly harm humans, they still pose a risk to food security and can have a significant negative economic impact on fisheries through decreased landings, decreased product marketability, increased aquaculture mortality or higher culture costs.

Figure 5.4. Epizootic shell disease.

Source: Maynard, Jeffrey & van Hooidonk, Ruben & Harvell, Catherine & Eakin, C. Mark & Liu, Gang & Willis, Bette & Williams, Gareth & Groner, Maya & Dobson, Andrew & Heron, Scott & Glenn, Robert & Reardon, Kathleen & Shields, Jeffrey. (2016). Improving marine disease surveillance through sea temperature monitoring, outlooks and projections. Philosophical Transactions of The Royal Society B Biological Sciences. 371. 20150208. 10.1098/rstb.2015.0208.

5.3 METHODOLOGY

The foundation of approved methodologies like clinical examinations and electron microscopic examination of tissue; however, in many of these situations, a summary of healthy histology is not available. In order to identify new and existing pathogens and parasites, more sensitive techniques include exploratory metagenomics, bioassays, in situ hybridization (ISH), conventional and real-time quantitative polymerase chain reaction (PCR and qPCR) as well as molecular diagnostic techniques. The risk and possible impact provided by a parasite can then be determined by having a thorough grasp of parasite transmission processes and host-related impacts. Histology remains the gold standard for undertaking a generalized screening of host tissues to detect the presence of parasites and their accompanying diseases, despite the growing accessibility of innovative molecular methods. To identify latent or subclinical infections, these techniques are then supplemented by in situ hybridization (ISH) or other molecular techniques. The presence of a specific parasite is frequently detected by readily accessible molecular diagnostics, but these tests frequently fail to provide useful data on viability, tissue tropism or the spatial variation of expression patterns linked to host immunity and parasite localization.

5.4 PATHOLOGY

Histology is frequently employed as a baseline search tool to identify and diagnose disease. Histopathological analysis is subjective, and many publications that describe new parasites don't adequately depict healthy or normal tissues. Histological atlases of healthy tissues aid in the identification of anomalies and improve researchers' ability to spot disease, stress reactions and the pathological effects of other non-infectious external causes. Fewer modern writings offer photographs of healthy tissue or a description of a baseline for researchers to compare, while more current works provide in-depth descriptions of infection in the histological section. When trying to come to a consensus among pathologists on a particular lesion or parasite, defining normal tissues is crucial. However, for many crustaceans, this is not possible because there are few documented criteria for comparison. Prior to recent reports, it was believed that lobsters were less likely than other decapod crustaceans to be parasitized. The fishing and aquaculture sectors target Palinuridae, which can both put lobsters under more stress and make disease transmission easier. The *Panulirus argus* Virus 1 (PaV1), the only virus known to exist naturally in a lobster species, is found in the family

Palinuridae (Li et al., 2006). Protozoan parasites, in particular ciliates, are typically thought to be the cause of significant crustacean diseases. They can result in post-harvest losses of American lobsters (*Homarus americanus*) of up to 15% each year. The disease was initially identified as a pathogen of captive lobsters kept in flow-through systems at low temperatures in Eastern Canada. Since then, it has caused epizootics in eastern North America and is now regarded as one of the three most significant infectious diseases of American lobsters. The majority of pathological studies of ciliate infections in invertebrates refer to crab hosts. The sole pathological investigation is into the emergence of protozoan disease in crabs raised under artificial conditions. There are few, inconsistent descriptions of the lesions that develop during spontaneous disease outbreaks in lobsters. There haven't been many studies done on experimentally caused disease in lobsters. The creation of experimental models for investigating lobster health and infectious diseases would enable researchers to compare their findings and evaluate the effects of infectious diseases on populations of wild and domesticated crustaceans.

5.5 VIRUS

The causative organism has been identified as the white spot syndrome virus (WSSV). White spot syndrome is one of the most devastating viral infections of shrimp (Sánchez-Paz, 2010). In India's shrimp culture businesses, this virus has resulted in losses of several million dollars and the losses are still being incurred. Its morphology was examined when the WSSV was isolated from *Penaeus monodon*. Crabs, copepods and other arthropods are among the vast variety of hosts for WSSV, which has been proven to be extremely harmful to penaeid shrimp. 43 species of arthropods, either from culture facilities, the wild or experimental infection, have been identified as hosts or carriers of the WSSV. The penaeid shrimp *P. monodon*, *P. indicus*, *P. japonicus*, *P. chinensis*, *P. penicillatus*, *P. semisulcatus*, *P. aztecus*, *P. vannamei*, *P. merguiensis*, *P. duorarum*, *P. stylirostris*, *Trachypenaeus curvirostris* and *Metapenaeus ensis*, the caridean shrimp were potential host for WSSV. WSSV has also been detected by PCR, in situ hybridization or immunological techniques in the crabs *Scylla serrata*, *Portunus pelagicus*, *Portunu sanguinolentus*, *Charybdis* sp., *Helice tridens*, Planktonic copepods and pupae of an Ephydridae insect have been found to be susceptible to WSSV. However *M. rosenbergii* has been found to be resistant, according to studies on the disease's pathogenicity for juveniles and adults of these three species. Freshwater crabs, 20 species of marine crabs, and *Artemia* sp. have all been shown to exhibit experimental WSSV infection. PCR, Western blot and histological analyses on the target organs of WSSV were done in these lobsters to see if they are the same as those in marine shrimp and to rule out the idea that they could serve as reservoirs or carriers of WSSV for marine shrimp.

In North American coastal areas, the American lobster *Homarus americanus* plays a significant economic role. Even though the American lobster fishery has historically been large and still is, there haven't been many contemporary studies on health and disease. Evaluating viral and non-infectious diseases in maritime environments presents significant obstacles. Factors in Disease Outbreaks; Development of Forecasting Models of Disease in Aquatic Hosts, Sensitive to Environmental Understanding the etiology and ecology of marine diseases can be significantly hampered by these problems. Surprisingly, American lobsters have comparatively few diseases. Gaffkemia, shell and bumper car disease or BCD may play significant roles in the natural mortality of lobsters. The complexity of these diseases' breakouts serves as a stark reminder of how dynamic the interior and external settings of lobsters are. One of the most significant and well-known infectious diseases of American lobsters, gaffkemia primarily affects lobsters that have been impounded. *Aerococcus viridans (var.) homari*, the etiological agent, is a free-living, gram-positive, tetrad-forming coccus that causes systemic disease in homarid lobsters. Because the bacterium cannot pass through an intact integument of lobster and cannot withstand stomach acids, the host is invaded through wounds or cuticle punctures. In its free-living stage, the organism thrives outside of the host in the benthic environment and in storage spaces.

5.6 SURVIVAL

The survival of lobsters and changes in total hemocyte counts during the course of gaffkemia appear to be influenced by a number of variables within the triad of host, pathogen, and environment. Bacterial strains, handling, temperature changes and trauma are some examples of stressors. Homarid lobsters are the most prone to infection and disease, despite the fact that there are decapod reservoirs of *A. viridans* infection in the marine environment. Clinically, infected lobsters have no obvious symptoms other than weakness or lethargic behavior and a spread-eagle posture that appear in the latter stages of the disease but are not pathognomonic for gaffkemia. Following infection, lobsters quickly develop anorexia. The pathogen only emerges in the hemolymph much later in the infection, after the bacteria have multiplied rapidly in the hepatopancreas and heart. The growth medium for *A. viridians*, however, is hemolymph. Red tail disease is so named because it can cause the ventral abdomen and hemolymph to turn pink. With no negative impact on the organisms, the majority of bacteria proliferate easily in fixed and moving phagocytes. No bacterial toxins are created. Death is most likely caused by metabolic impairment brought on by hepatopancreatic dysfunction. Additionally, there is a pronounced hemocytopenia and dysfunction in the clotting system.

An inflammatory-like reaction manifests during the course of the disease, according to the pattern of total hemocyte counts. Several culture techniques, such as the presumptive phenylethyl alcohol broth test, and hemolymph smear examination, can be used to identify early infections with A. *viridans* in lobsters. Culture techniques normally need 4–7 days (Stewart et al., 2011). Heavy infections can be identified by the co-agglutination technique of Saxegaard and Hastein or a direct Gram stain of hemolymph smears. A polymerase chain reaction/oligonucleotide probe that is very sensitive and specific for the detection of *A. viridans* is one of the more recent diagnostic methods. But none of the examined strains belonged to *A. viridans var. homari*. Investigations into human diseases typically use this relatively quick (5 h) approach. The development of an indirect fluorescent antibody method to identify *A. viridans* in lobsters is a significant development. *A. var. homari* strains were used to make the antiserum. The procedure is precise and quick: for light infections that only need bacterial incubation, the time needed is only 48 hours; for heavy infections, the time needed is only 2 hours when applied directly to hemolymph smears. A significant number of (>1000) lobsters with a presumed diagnosis of gaffkemia were used to critically assess the methodology using conventional bacteriological methods. Large-scale surveys focusing on *A. viridans* in populations of wild lobsters can be quite useful when using this strategy.

The sole medication permitted for use in Canada is oxytetracycline, despite the fact that *A. viridans var homari* is susceptible to a number of antibiotics and disinfectants. To prevent and treat gaffkemia in captive lobsters, the product is added to the meal. However, sick lobsters are anorexic and unlikely to consume enough medicated feed during disease outbreaks. At 40°C, the withdrawal time is at least 30 days long. The persistence of oxytetracycline in sediments, bioaccumulation by bivalves or other crustaceans, and bacterial resistance development are the main issues with its use in the marine environment. Human sensitivity to oxytetracycline is also a problem; the more the public views and accepts a fishery and its products as "green," the better. Although outdoor testing was recommended, numerous disinfectants were successful in experimental settings. The efficacy of modern disinfectants should also be considered for usage in lobster holding facilities. In laboratory trials, formalin-killed bacteria were used to create low levels of protection, while a combination of vancomycin and live bacteria was used to induce high levels. The efficiency of these kinds of vaccinations was further demonstrated) in both laboratory and field studies, including the efficacy of immersion immunization. Significant improvements over older steam-sterilized vaccines include ease of preparation and the absence of the need for vancomycin to produce an immunogen that is effective. A significant factor in determining the level of protection brought on by vaccination was the stage of the molting cycle. Lobsters caught in the spring, which had recovered more fully from the stress of ecdysis, were better protected than those caught in the late fall or early winter, which had recovered less.

A. viridans var. homari strains vary in virulence, as evidenced by observations of lobster survival following both natural and artificial infections with the bacterium. The bacterium's ability to form capsules, susceptibility to lobster agglutinin and ability to proliferate in lobster serum all decrease with loss of virulence. Virulence variations most likely reflect alterations in gene expression brought on by environmental conditions that result in antigenic variation. The molecular genetics and epidemiological investigations of *A. viridans var. homari* should also find utility for random amplified polymorphic DNA analysis. American lobsters are frequently infected with and afflicted with gaffkemia. The effects of gaffkemia on the wild fishery are unknown, despite the fact that it is acknowledged as a significant disease of lobsters. Early lesions are bilaterally symmetrical and center on the dorsal carapace's setal pores. As the disease worsens, the round, black lesions overlap and gather together. Vibrio-related bacteria are found in and on lesions. Winter harvesting of lobsters is linked to this type of shell disease; the importance of overcrowding, poor water quality, and inadequate foods in disease progression and severity may be significant.

5.7 LESIONS

Lesions are distinct, round, black and present in a variety of spots on the lobster, starting at the pores and pits of the carapace, in burnt (rust) spot shell disease. This morphology is frequently connected to pollution and is seen in lobsters in offshore canyons. The lesions contain a variety of fungi and bacteria. Although there is no definitive connection between burnt spots and disease, it is probable that burnt spots are an early stage of the disease. A recently discovered disease linked to global climate change may be epidemic shell disease. Host, pathogen and environment interactions and connections play a variety of roles in determining the prevalence and severity of the disease. The leaning response of lobsters may be negatively impacted by thermal stresses, which are visible in oceanic warming on local, regional and global scales, predisposing them to shell disease. Epidemic shell disease is not easily contagious from one lobster to another, just like other types of shell disease. In order to start a disease, a pathogen, host factors and one or more undiscovered environmental variables must all be present. For an infection to start, bacteria from the family Flavobacteriaceae need to be on the lobster's shell. Instead of chitin, environmental stressors or poor nutrition may be to blame for this cuticular aberration, which is produced by altered carbonic anhydrase deposition.

Season, sex, and size of the investigated lobsters all affect the prevalence of epidemic shell disease. Prevalence is generally highest in May and June and lowest in August, when lobsters have molded into new shells. All lobster life stages, including Stage III larvae, can develop the disease, including larvae, young of the year and adults. In younger lobsters, frequent molting presumably lowers the incidence and severity of disease or the start of shell disease. Since spiny or slipper lobsters harden their shells quickly, this type of shell disease does not affect them. Small lobster tends to molt a lot, but American lobster may also harden at varying rates, which could result in selection for disease resistance. Large, ovigerous female lobsters are more susceptible to epidemic shell disease and their intermolt period is longer than that of male lobsters of comparable size.

5.8 SHELL DISEASE

The likelihood of developing shell disease in lobsters may be significantly predicted by their origin. Southwest Nova Scotia lobsters are more likely to have impound shell disease than lobsters caught locally, regardless of whether they are impounded in Maine or New Brunswick (Barnett & Anderies, 2014). This could be due to decreased hemolymph levels, decreased total hemocyte counts, increased metal contamination levels, and decreased general health and nutritional status at the time of impoundment. It is currently unknown, though, if any of the three varieties of shell disease susceptibility has a hereditary component. The health of lobsters and their susceptibility to viruses or other stressors may both be strongly impacted by their diet. Commercial Artemia diets have been shown to affect the growth, survival, and health of young, nutritionally sensitive benthic stage lobsters up to the age of two. On the surface of lobster shells, a complex community of microbes including amoebae, algae, bacteria and other protists dwells in biofilms.

The lack of a laboratory model to examine the infection and disease in lobsters is a major obstacle to the study of epidemic shell disease. Shell disease is not very communicable, is challenging to duplicate in both the lab and the field, and requires a trigger to manifest. As previously said, the development of disease requires the interaction of a pathogen, a host and an environmental element. The lobster carapace must contain strains of *Pseudoalteromonas gracilis* and members of the lobster Flavobacteriaceae species complex as a prerequisite to infection and the subsequent development of epidemic shell disease. Although their impact on the pathogenesis of shell disease has not been examined, it was hypothesized that bacteriophages may greatly influence bacterial virulence and evolution. In the intricate, dynamic internal and external habitats of lobsters, there are complex microbial relationships interacting. Surprisingly, it doesn't seem like pollution affects where and how severe the outbreak of shell disease is. Between 2000 and 2003, the prevalence of epidemic shell disease was highest in New York in areas of the water with the least pollution, such as eastern Long Island and the south shore of Long Island, and lowest in areas of the ocean with the most pollution. Sewage contamination is no longer seen as a major contributor to the development of shell disease in an ongoing research study of shell disease in edible crabs *Cancer pagurus* in areas where water quality has greatly improved.

5.9 ENVIRONMENTAL FACTORS

Environmental factors have a significant role in the dynamics of host-pathogen-environment interactions that result in disease. Larger populations or those in nearby ecosystems may be impacted by these interactions. Further research is necessary, especially across wide geographic areas, to determine variations in food and pollutant levels in impacted and unaffected lobsters. Currently, it is unknown what part environmental pollutants have in the epidemiology of epidemic shell disease. There are four main categories of contaminants: (1) metals, which can have negative health effects on lobsters; (2) pesticides, which can also have negative health effects on lobsters; (3) brominated flame retardants, which may or may not have a negative impact on lobster health; and (4) phthalate esters, which may or may not have a negative impact on lobster health. These contaminants can change cuticle formation, which increases the likelihood of bacterial invasion, alter neurological activity, decreases cleaning behaviors, change endocrine activity, , damage gills and respiratory activityand alter gonad genetics.

The latter have negative impact on population numbers and structures by reducing reproductive fitness. The marine environment along the north-eastern coast of North America contains high concentrations of these four pollutants. Alkyl phenols, strong endocrine disruptors in vertebrates, are also a major source of concern when it comes to the etiology of epidemic shell disease. These leftovers of industrial processes and the organic breakdown of plants are frequently present in the marine environment and are tenacious and challenging to eliminate. Alkylphenols are found in marine sediments, lobster hemolymph, and tissues. Due to greater amounts of the molting hormone ecdysone, lobsters with shell disease molt more frequently than lobsters without it. Ecdysone levels are extremely high in ovigerous female lobsters with shell disease. Perhaps lobsters' molting serves as a defense mechanism against shell disease. High amounts of alkylphenols may interfere with shell development and hardening, leaving lobster shells more vulnerable to bacterial invasion and eventual bacterial destruction. By soaking in clear water for an undetermined amount of time, lobsters can reportedly become free of alkylphenols. Perhaps levels of alkylphenols should be checked throughout the lobster's life cycle. The recent outbreak of epidemic shell disease in New England lobsters may have its roots in temperature as the main environmental component. The prevalence of shell disease increased southward along the coast of Massachusetts. Shell disease may affect lobster movements or habits in addition to cosmetic effects on shells, accelerated molting and premature molting of ovigerous females. Behavioral thermoregulation of lobsters to avoid temperature extremes is not a significant factor in the progression of epidemic shell disease. Affected lobsters' mobility behaviors were not different from those of unaffected lobsters when it came to epidemic shell disease. Evidently, lobster with pandemic shell disease behaves more like a shelter animal. This might have a negative effect on interactions between conspecifics, including behaviors related to feeding and reproducing as well as mortality and molting rates.

Epidemic shell disease is a big problem in the New England lobster fishery even while impoundment shell disease and rust spot shell disease of lobsters have traditionally occurred at relatively low prevalence levels. At the moment, the success of the American lobster fishery depends on the use of fish as bait, and lobsters eat a lot of bait. In an experiment, lobsters fed fish-only diets either developed shell disease or perished; those offered mixed diets had higher survival rates and less shell disease (Tlusty et al., 2008). Fish diets put lobsters under nutritional stress and lower their health status, which causes more disease and higher mortality. The bacteria Aquimarina were recently shown to be present in all lesions of impoundment and pandemic shell disease in lobsters. Along with the multifaceted character of shell disease, sick lobsters also lack immunity and have weaker exoskeleton defenses. There are a lot of unanswered concerns regarding the ecology of the lobster population, how shell disease interacts with it and how to manage disease consequences on lobster fisheries. Although there are many more bacterial genera and species in marine habitats, there aren't many other infectious diseases that have been reported. In lobsters kept in artificial sea water at 20–22 °C and in (natural) sea water at 23 °C, specific Vibrio species were first linked to disease. These lobsters were also probably under heat stress due to temperature (Bowser et al., 1981). Uncertainty exists over whether these Vibrio species are primary infections or secondary to different stressors on lobsters.

The bacteria did not grow at temperatures above 23 ° C, thus it is unlikely that this group of *V. fluvialis*-like strains are human pathogens. There are no medications authorized to treat this condition. It has not been established whether additional pathogenic marine Vibrio species cause similar sickness in lobsters. The outbreak may have started in impounded lobsters and spread to wild lobsters as a result of global climate consequences. The epidemiology of Limp Lobster Disease, however, is still a subject of much debate. The filamentous bacterium *Leucothrix mucor* may induce fouling of eggs in hatchery facilities with inadequate husbandry i.e., crowding, poor water quality and flow as well as inadequate nutrition. Bacterial fouling on lobster eggs can be controlled by antibiotic treatment (L.H. & Jones, 1999). Fungus-like protists are diseases of crustaceans, usually in facilities used for aquaculture. Species of Lagenidium can infect and then cause disease in embryos and larval stages. Affected eggs will not hatch and depending on their developmental stage, larval lobsters will survive for varied amounts of time. The longer the period of survival, the thicker is the cuticle. Malachite green and fungicides can be used to disinfect culture systems. Because *Alteromonas* sp., a Gram-negative bacterium, covers healthy eggs of *H. americanus* and other crustaceans, they are quite resistant to infection with *L. callinectes*. The amount of this protective bacterium on eggs may decrease under unfavorable circumstances in hatcheries and culture facilities.

In culture facilities around the world, the water mold *Haliphthoros mildfordensis* has been shown to kill eggs and larval crustaceans. Poor husbandry practices contribute to disease, and better husbandry practices will reduce health issues. Malachite green to antifungal chemicals are just few of the available therapies and disinfectants. On the other hand, it is fervently argued that adequate husbandry, such as cleaning and prevention, was significantly more cost-effective and preferable over therapy because infected lobsters would have a bad prognosis due to their weakened state. Both warm and cold water farmed crustaceans may develop black, shell disease-like lesions as a result of the fungus Fusarium. This common fungus presents a problem since it is an opportunistic infection that attacks both young and mature lobsters through dead or damaged tissues. Bacterial infections may spread laterally.

Unfortunately, except for proper husbandry, there are no practical or efficient control measures for Fusarium disease in lobster facilities. The common marine amoeba *Neoparamoeba pemaquidensis* is a serious disease of salmonid and non-salmonid fish hosts as well as marine invertebrates, including American lobsters. Its impacts are exacerbated by a confluence of environmental and anthropogenic variables, which suppress the lobsters' natural defenses and made it possible for the opportunistic amoeba to infect them. Koch's hypotheses were not supported, despite the fact that the parasite may spread from infected to uninfected lobsters through cohabitation. Unfortunately, improved coordination and communication between experts and business stakeholders are required to ascertain different facets of BCD epidemiology in the lobster fishery. When the water temperature is chilly, an outbreak of BCD can occasionally be seen in impoundment facilities. This thorough investigation covered gross and microscopic pathology as well as the successful small subunit rRNA gene sequencing of *A. hemophilia*. The investigation showed that isolates are persistent both spatially and temporally. An alternative argument is that infections associated with lobsters may travel between locations naturally or as a result of industry transit of lobsters. Similar to the description of ciliate-induced lesions in prior experimental research, the natural disease progressed in two phases and included necrotizing bacterial enteropathy and hepatic pancreatopathy.

CHAPTER 6

Climate Change and Lobsters

CONTENTS

6.1 INTRODUCTION

The rapid loss of marine populations over the past 30 years is primarily due to habitat deterioration, overfishing and climate change. One billion people, largely in developing nations, rely only on fish as their main source of protein. Fishing and aquaculture provide jobs for 14% of the global population and provide $2.8 trillion in economic benefits annually ("Warming Pushes Lobsters and Other Species to Seek Cooler Homes," 2019). However, more needs to be done to comprehend and plan for how climate change may affect global fisheries and marine ecosystems. The collapse of ground fish stocks caused this sector to expand in some regions (Figure 6.1).

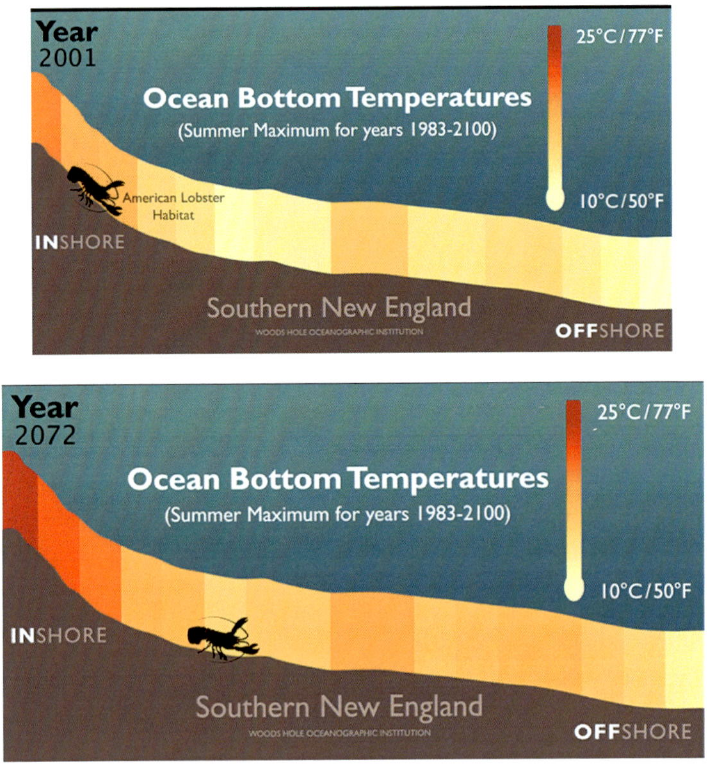

Figure 6.1. Climate change and lobsters.

Source:https://blogs.agu.org/geospace/files/2018/01/FINAL-1.gif

Coastal lobsters are a highly sought seafood delicacy all over the world. But lobsters, especially spiny lobsters, are particularly vulnerable to climate change because of their lengthy larval stage. Indeed, warming oceans have been a primary factor in the documented consequences of climate change in a number of lobster species around the world. The influence of climate change on biodiversity has been extensively studied using ecological niche models. These models extrapolate a particular bioclimatic envelope using distribution information from several species together with environmental factors. By estimating prospective changes in the habitat appropriateness of the species under study under various climate scenarios, it is possible to draw conclusions about potential climate change implications (Greenan et al., 2019).

6.2 DIVERSITY

The presence of diverse clades with tropical affinities, but occurring at lower abundances when compared with those present in temperate waters, strongly influenced the projected pattern for higher diversity towards the tropics. Coastal lobster species were anticipated to lose between 40 and 100 percent of their habitat, with the majority of losses occurring in the tropical zone. Greater losses of spiny lobsters are anticipated throughout the shores of eastern Africa, the Indo-Pacific region, and the wider Caribbean region/ Brazil. These losses are caused by the anticipated changes for the genus *Panulirus* and will probably have an impact on the economies of the affected nations because countries like Australia, which produces the fourth-highest volume of lobster in the world, and Indonesia, which produces the fifth-highest volume; 98% of which came from the capture of *Panulirus* spp. The top 10 countries that only harvest *Panulirus* spp. are Brazil, the Bahamas (*P. argus*) and Nigeria (Faria-Junior, Gaeta, & Freire, 2013). Even though it has fewer species, the genus *Jasus* is expected to see major range contractions in regions where lucrative fisheries are focused on this resource, such as the coasts of South Africa (*J. lalandii*), South Australia, and New Zealand (*J. edwardsii*). However, the cultivation of *P. homarus* and *P. ornatus* is flourishing in the Indo-Pacific region, bringing in USD $31,519 million in 2016. It is essential to take into account anticipated changes in habitat appropriateness while creating regional studies for stock management, creating new aquaculture enterprises, and creating protected areas.

H. americanus, *H. gammarus*, and *N. norvegicus* account for 74% of the world's lobster production, making clawed lobsters far less diverse but nevertheless extremely profitable (Pere et al., 2019). Large commercial fisheries are focused on both genera, which are only found in temperate waters in the Northern hemisphere. According to models, all species are expected to suffer a move northward and lose their HS in the southern area.

The North European, North American and Canadian fisheries will probably be impacted by the expected envelope shifts and loss of suitable habitat, which could have a negative impact on the livelihood of coastal people. However, recent research employing regional models with finer spatial resolution revealed that the Gulf of Maine region experienced geographic variation in the degree of temperature effects on species distribution. The data reported here should be taken into consideration at the coarse-scale they were created at because these local or specific particularities are difficult to represent in a worldwide study. Additionally, it is crucial to note that even though predictions of decreased habitat suitability may appear to conflict with those of other studies, they can be easily explained by the use of climatic variables that are typically excluded from other studies, such as dissolved oxygen concentration at the surface and ocean surface pH. The relationship between temperature and species distribution has long been known, but other climatic conditions, such as ocean acidification, have also been shown to be harmful, especially for crustaceans like lobsters that rely on carbonate to form their shells. The loss of suitable habitat was thus expected to occur at higher latitudes as well. In reality, pH proved to be a key driver in the distribution of coastal lobster species.

With increases in water temperature, modifications to ocean currents, pH levels, and winds, climate change is expected to have an impact on the world's oceans. Lobster stocks are particularly susceptible to climate change because of their lengthy larval life. Australia's western and southern rock lobsters are impacted by long-term temperature increases in the sea, which are expected to persist. The size of maturing western rock lobsters and their migration have been affected by these increases. The decrease in lobster abundance and rise in lobster diseases in the southern portion of the *Homarus americanus* population as well as the increase in captures in northern parts have both been linked to climate change. Changes in the number of Japanese and Hawaiian spiny lobsters have been linked to a north Pacific regime shift. Monitoring pre-recruits becomes crucial in this situation, and management techniques must be attentive to patterns in recruiting abundance. There is a necessity to scale down climate change models to a scale that is relevant to lobster stocks both spatially and temporally as well as the uncertainties associated with climate change projections and their impact on the ecosystem. The effects of climate change are introducing greater pressure on the world's already fragile fisheries. It is important to look at the effectiveness of risk perception as a tool to inform how to convey the science of climate change and suggestions for management in connection to the creation of adaptation plans for fisheries. The majority of fishers surveyed expressed skepticism about whether climate change was a real process despite having observed similar significant changes in the maritime environment and lobster industry that are consistent with climate change. The crucial aspect of the industry's adaptation to climate change is fishers' perceptions of risk to make it tough for them to acknowledge the problem of climate change. As a result, there is a barrier to future action on the topic of climate change as well as to communication and awareness about the problem. The creation of communications that are acceptable to fishers on a cultural level will be necessary to surge their understanding of climate change and, consequently, their capacity to adapt.

6.3 ECTOTHERMS

The abilities of ectotherms to molt and grow are normally enhanced by warmth, but extreme warming can impair these abilities and even result in death, which has an impact on their demography. *Homarus americanus*, an American lobster, has been the subject of research on the effects of warming on benthic adults, but the performance of planktonic larvae at exceptionally high temperatures has received little consideration (Harrington et al., 2019). The physiological performance of larvae may be compromised earlier than that of adults due to warming since water column temperatures are predicted to rise more quickly than bottom temperatures. In order to accurately estimate the implications of climate change, it is crucial to define the stressful and deadly temperatures for lobster larvae. However, more ecologically significant longer-term high-temperature exposures lasting days or weeks have rarely been tested for larvae of this species. Mortality following short-term exposure to 28–36 °C from minutes to hours in duration has been documented. Short and long-term exposures to temperatures between 20 and 26 °C were observed to cause sublethal stress markers, such as irregular heartbeat, which may indicate that prolonged exposure to these temperatures could be fatal. If temperatures up to 30 °C are actually stressful or fatal to larvae over the next 10 to 50 years, this could have a significant effect on lobster populations. The necessity for more research into lobster larvae's resistance to high temperatures is highlighted by the knowledge gaps that have been uncovered.

6.4 MARKET

Coastal towns and the seafood market will likely see socioeconomic effects from climate change's effects on fisheries. However, it is difficult to incorporate climate change data in a way that planners for adaptation, decision-makers and fisheries managers can use effectively. Researchers use a spatial viewpoint to analyze the effects of climate change on the dominant fishery, in this case, American lobster in Nova Scotia, Canada. Economic reliance on the fishery, population size, the variety of fishery revenue, the condition of harbor infrastructure, the total cost of replacing each harbor, rising relative sea levels and flooding, the vulnerability of offshore lobster to ocean warming and changes in zooplankton composition and anticipatory changes in fishery productivity across management borders are all factors taken into account.

Figure 6.2. *Chionoecetes opilio.*

Source:https://upload.wikimedia.org/wikipedia/commons/2/2e/Chionoecetes_opilio_Kanikkokan.jpg

hanges in offshore lobster distribution are predicted to have a neutral or positive impact on the region as a whole using two ocean models to generate multi-decadal scale estimates of bottom temperature. However, it becomes clear that adaptation planning is required for long-term sustainability when lobster sensitivity is joined with climate change-related vulnerabilities of coastal fishing communities. Planners for climate change adaptation and managers of fisheries can use this impact assessment as a framework and source of data for more in-depth analysis. In line with the global trend of rising ocean heat content brought on by climate change, ocean temperatures on the continental shelf off the Northeast USA and southern Atlantic Canada have grown during the past 50 years (Kavanaugh et al., 2017). The resultant biological effects differ by species and place. As global temperatures rise, certain species may find it impossible to survive, while others may thrive. If waterways become unsuitable for species of economic importance and range shifts result in a decline in local abundance, these long-term changes could have a significant influence on commercial fisheries.

The most profitable fishery in Canada is the American lobster (*Homarus americanus*), which provided 44% of the commercial value of all fisheries in Atlantic Canada in 2016 (Rochette et al., 2018). Although the snow crab (*Chionoecetes opilio*) (Figure 6.2) is significant in the northeast, many rural towns in Atlantic Canada rely significantly on lobster for their economic survival. Due to stress, decreased recruitment and increased disease, ocean temperatures above an ideal thermal range can hinder lobster survival, development and reproduction. Within its region, lobster response to warming varies in size and characteristics. In the Gulf of Maine, where the lobster business has first benefited from the loss of lobsters' natural predators, warmer bottom temperatures and strict conservation efforts, lobster numbers reached a record high. In contrast, lobster populations are dropping in southern New England's warmer climate, where there are fewer protective measures.

With the potential to have large economic effects, ocean warming may cause changes in commercially important species. Nowhere are those effects more apparent than in the Gulf of Maine, home of the American lobster, the most valuable marine species in North America. However, as monitored shallow water lobsters fall and landings shift to the northeast, there are rising uncertainties that area maritime industries would suffer. A more positive aspect of climate change is produced by the interaction of ocean warming, tidal mixing, and larval behavior. A complicated interaction between the settling behavior of lobster larvae, climate change and regional oceanographic circumstances led to this rise. Post larval sounding behavior is restricted to a thermal range between 12°C and 20°C in particular. The thermal envelope for settlement has typically included summer thermally stratified surface waters in southwest regions. The steep depth-wise temperature difference made thermally appropriate locations for larval colonization to expand only moderately, despite the fact that surface layers are warming here the fastest. In contrast, the northeast inhibits thermal stratification due to significant tidal mixing, and recent ocean warming has rendered a large portion of the seafloor more conducive to larval colonization. The larger amount of thermally appropriate habitat brought on by warmer summers is correlated with the recent declines in lobster settlement numbers seen at shallow monitoring sites. This leads to speculation that the increased area of suitable habitat may contribute to the explanation of the significant gains in lobster population observed in this area over the past ten years and counteract any potential future decreases. It also implies that comprehension of the local interactions between biological thresholds that are particular to life stages and smaller-scale oceanographic processes is necessary to predict how fisheries would fare in a changing climate.

6.5 CLIMATE CHANGE

Global climate change has a lot of information but appears to be difficult. The majority of the world's seas are warming, but patterns in the distribution, quantity and rate of warming vary through time and geography and in ways that can have a significant impact on marine life. The climate velocity of the net poleward shifts in the distribution of marine organisms that are commercially fished or ecologically significant to marine ecosystems has been documented in numerous studies. However, understanding how those features of climate change impact species where they reside is necessary to evaluate the implications of ocean warming. The majority of studies use sea surface temperature (SST) trends as a practical and significant indicator of regional warming, but these remote sensing metrics might not accurately reflect exposure for organisms whose crucial life history processes take place on or near the seabed. The thermal characteristics of the sea's temperature substantially vary across time and space in the northwest Atlantic's coastal and shelf waters. Thermal thresholds have an immediate effect on the demographic and fitness processes that affect species. The 12°C threshold for lobsters represents a threshold below which colder temperatures have lethal and sublethal behavioral avoidance and decreased size at molt effects (Rochette et al., 2018). The decoupling of larval supply from settlement patterns is caused by small temperature changes around this threshold that are accompanied by substantial variations in survival and settlement. As a result, larval transit and survival may be hampered by the 12°C threshold in an ecological sense. Therefore, ocean warming has the potential to have a considerable impact on the survival and dispersal of lobster larvae. This is particularly true in the Gulf of Maine, where forecasts for the end of the century indicate an increased area and length of time the seafloor spends over the thermal threshold.

This modifies how we interpret decades of data on young-of-year (YoY) settlement, which has steadily decreased for about a decade in the Gulf of Maine and presents a new picture of how ocean warming may affect the range and abundance of the American lobster. The question at hand is whether the most valuable fishery in North America is seeing a decrease. Western rock lobster (*Panulirus cygnus*) fishing puerulus settling is substantially impacted by environmental factors such the Leeuwin Current, which is influenced by the El Nino-Southern Oscillation cycle and westerly winds in late winter and early spring. Seasonally changeable water temperatures are rising, westerly winds in winter are diminishing and El Nino events are occurring frequently as a result of climate change. Increases in undersized and legal-sized lobster abundance in deep water relative to shallow water, variations in catch patterns and a drop in size at maturity and size of migrating lobsters from shallow to deep water are all possible effects of rising water temperatures during the past 35 years. The water temperature around the time of puerulus settlement influences the size of migrating lobsters. Evaluation of the impacts of climate change on puerulus settling, catchability, molting of females from setose to non-setose, timing of molts and peak catch rates is also needed. These biological changes are anticipated to continue as long as climate change models predict that the warming trend will persist. The effects of the changes, which must be taken into account in stock assessments and management, may be either detrimental or favorable. One of the most profitable American lobster (*Homarus americanus*) fisheries in the northeastern United States has historically been based in southern New England. Current years have seen a sharp reduction in the number of lobsters in the area, which has been accompanied by an increase in summer temperatures that are stressful and shell disease. Fisheries in coastal and shelf waters around the world that are commercially important are being directly impacted by climate change. This is particularly true of the North Atlantic, where ocean warming has surpassed that of other regions of the world, leading to altered phenology and regional range shifts for a variety of marine fish and invertebrates with significant implications for the management of fisheries in the future. Over the past 20 years, the American lobster population, *Homarus americanus*, has undergone significant alterations in southern New England. These modifications imply that the summertime climate in coastal regions is changing, making it unsuitable for this iconic, historically plentiful and commercially significant species. Southern New England experienced historically high levels of lobster production and abundance in the middle of the 1990s (Howell, 2012). The southern New England lobster stock is currently experiencing recruitment failure, according to the Atlantic States Marine Fisheries Commission's (ASMFC) Lobster Technical Committee. In contrast, sudden warming events have led to changes in phenology further north, where milder temperatures are more common. In the most recent instance, a record-breaking ocean heat wave in 2012 caused an early molt in the Gulf of Maine, resulting in an abundance of landed lobsters that overwhelmed processors and significantly lowered the price.

The winter-spring diatom bloom has virtually disappeared in the Narragansett Bay and nearby coastal waters, eelgrass has been declining, comb jellies have increased, the composition of the groundfish assemblage has changed. Summer water temperatures in southern New England have historically exceeded a long-recognized 20°C physiological threshold, which is particularly relevant to lobster. Overall, this causes the temperature niche of the lobster to shift northward, which means that in the next decades, coastal southern New England won't be a favorable nursery for the American lobster. Comparing current distributions to historical baselines is therefore useful. In terms of both production value and export value, rock lobster fisheries are Australia's most valuable wild fishery. Different lobster species are harvested and exported by different states, with the Hong Kong market receiving the majority of the landings. The different species are thought to be independent of the export market in the Australian lobster industry, therefore a change in landings of one species has no effect on the price of the others. The market integration of Australian exports to Hong Kong for the four species and several exporting states is examined. The linked structure of the Australian lobster export market to Hong Kong shows that, at least from an economic standpoint, the potential effects of alternative fisheries management and development methods at the state and species levels cannot be addressed in isolation (Norman-López et al., 2014). Additionally, all Australian lobster fisheries can be anticipated to be impacted by the effects of external shocks influencing productivity in one state, such as climate change.

Four different varieties of lobster are produced and exported by Australia and they come from various parts of the nation. Western rock lobster (*Panulirus cygnus*) in WA, Tropical rock lobster (*Panulirus ornatus*) in northern (tropical) Queensland (QLD) and the Torres Strait and Eastern rock lobster (*Jasus verreauxi*) in New South Wales are all harvested as rock lobsters. The Southern rock lobster (*Jasus edwardsii*) is caught in Tasmania (TAS), Victoria (VIC), South Australia (SA) and (NSW). More than 75% of the output value for each of these species was transported to China in 2010. They could optimize their supply to international markets and take into account the price impacts that climatic changes will have on their supply to global markets if they were aware of the extent of substitutability between the four Australian RL species and producer states. Producers will be better able to control pricing if they can coordinate supply if different Australian RL species are viewed as replacements by customers. Prices at the individual producer level will essentially be more flexible when coordinating since they together have more market power, indicating that equivalent supply reductions there will have a larger impact on prices and revenues. In the Chinese market, where Australia controls supply, this effect might be very substantial. In contrast, exports of several Australian RL species and/or producer states can be seen as discrete market sectors, where changes in supply by one do not affect the price or income of others and the effect on their own price will be less than their change in supply.

6.6 RELATIONSHIPS

Understanding the relationships between various producers and products can be done with the help of market integration analysis. When examining pricing correlations, the Johansen cointegration approach is frequently used in the agricultural and fisheries literature. This is significant because the price series must be non-stationary for this form of analysis and a single change in the trend's level or slope might make it harder to reject a false nonstationary null hypothesis. In addition to the stationarity characteristics of the series, this method takes into account non-stationarity tests that employ algorithms to determine the timing and frequency of breaks. It is important to take into account the economic significance and fit them into a multivariate cointegration model proposed by Johansen et al. that takes the breaks into account at specific moments in time. By employing this method, we take into account the series' statistical characteristics as well as and this is vital to our understanding of the pertinent historical events that led to breaks in the series. The investigation specifically focuses on the Hong Kong market because it is the biggest wholesale market for lobster in China and the largest market for live seafood in Asia. Since this comprises the majority of all Australian lobster exports, it is important to concentrate on live and whole fresh RL. More than 95% of the total exports of live and whole fresh RL from the tropical and eastern fisheries, 78% from the southern fishery and 57% from the western fishery go to the Hong Kong market. Using market integration analysis, it is crucial to look at the long-term link (substitutability) between the species exported by various nations. If there is a long-term link, we then think about how perfect or imperfect substitutability is, as well as how shocks affect the equilibrium relationship.

6.7 ANALYSIS

Research has focused on analyzing and describing specific variations in the abundance of commercial species over the past 20 years, as well as potential connections between ocean temperature and the size of fish stocks. The influence of climate change on fisheries is thought to be quite long-term in nature. However, recent years have seen descriptions of regime changes in the climate. The global ecosystems were significantly impacted by these changes. Many exploited communities' response to climate change appears to be a worldwide phenomenon. Variations in water temperature and rainfall have an impact on the dynamics of ocean currents, river flow and the number of wetlands they occupy. These changes can also have an impact on the productivity and distribution of marine and inland fisheries resources. This will have an effect on the distribution and production of fish populations as well as the structure and operation of the ecosystem. The safety and effectiveness of fishing operations are impacted by the increased frequency of extreme events, such as floods, droughts and storms. These occurrences also cause more harm and disruption to residences, infrastructure services and the coastal area. The ecosystem and way of life in coastal wetlands will be affected in an unforeseen way by rising sea levels, melting glaciers at the source of important rivers and other significant environmental changes.

Due to the intricate relationships between fishing, climate change and other industries, fishing has indirect effects. It will be obstructed by changes in the amount of water needed for agriculture, the cost and availability of food obtained from aquaculture and the redirection of financial resources from government and international fisheries management towards relief emergency efforts after catastrophic climatic disasters. The Caribbean's spiny lobster fishery is only second to shrimp in terms of economic significance. For many developing nations in the Caribbean region, these fisheries are also of significant social and even political importance (Monnereau & Pollnac, 2012). The Pan-Caribbean spiny lobster concept was made possible by the species' unusually lengthy lifespan, whose larvae can spend up to ten months in the strong currents that dominate the area. As a result, the products of spawning in upstream locations can result in settlements in down-river places. There is currently no regional framework for the control of fishing capacity, which has significant implications for the management of the various fisheries.

It is anticipated that human-induced climate change would raise ocean temperatures and alter circulation patterns, potentially having significant effects on the movement and survival of planktonic larvae of marine animals. Larval distribution is influenced by circulation, while survival and growth are influenced by temperature. But in a future climatic scenario, the combined impact of changes in circulation and temperature on larval dispersal and survival has rarely been investigated. Such knowledge is necessary to forecast future species distributions, foresee changes in ecosystems and create efficient management plans. Ocean warming favors lobster larvae survival, whereas a strengthened western boundary current reduces larvae supplies to the shore by limiting cross-current larval dispersal. Additionally, we can discover that changes in circulation have a greater impact on the connectivity patterns of lobster larvae along southern Australia than ocean warming in the future climate, resulting in a 4% decrease in larvae supply to the coast and a 270 km poleward shift in the settlement peak in the model simulation. One of the chief causes of changes in species distributions brought on by climate change may be ocean circulation.

The early spring rising of ocean temperatures and an associated increase in lobster landings during a marine heat wave in 2012 disturbed the fisheries. The supply system was not ready for the early and plentiful landings of lobsters, which caused a price fall. Researchers created a projection for the transition of the Maine (USA) lobster fishery into its high-volume summer landings phase as a result of this sequence of events. The forecast employs a regression method to connect the opening day of the fishery's high landings period to spring ocean temperatures obtained from four NERACOOS buoys off the coast of Maine. The forecast, which has been tested against past conditions, can predict the start day to within one week of the actual start and can be released three to four months before the start of the high-landings period, giving the fishery and its associated supply chain crucial lead time to get ready for the upcoming season. Users can evaluate the confidence and consistency of the forecast and take the uncertainty into account when using the information in a particular year because forecast results are communicated in a probabilistic fashion and updated weekly across a 6-week forecasting period. This kind of seasonal prediction focuses on the timing of events, giving users access to climate-relevant data at time scales that matter for operational choices. This kind of forecast can help fishing industry participants better adapt to and get ready for operating in the context of climate change because climate change modifies seasonal phenology and decreases the validity of past experience as a reference for future expectations.

Lobster fisheries in Australia are the most profitable in terms of both total production and export value. Along the majority of the continent's coastline, they also provide support for commercial, recreational and indigenous fishers. Based on biological traits, fishery data collection, assessment and management techniques and supply chain considerations, we can evaluate the significant similarities and variations across these lobster fisheries. Australia is home to a wide variety of palinurid lobsters, but only three genera, spread over eight different management jurisdictions, sustain substantial fisheries. Landings are dominated by catches of the western rock lobster, *Panulirus cygnus*, with the southern rock lobster, *Jasus edwardsii*, tropical lobster and eastern rock lobster. These fisheries are being affected by large-scale environmental factors like climate change in similar or distinct ways, necessitating new management and increasing the need for greater resilience in present supply networks. Despite the fact that these fisheries are distinct from one another, the integrated character of the main Chinese export markets points to potentially significant economic and market-related connections. Given that many fisheries involve indigenous Australians, approaches that also take socio-cultural factors into account are crucial.

6.8 EMISSIONS

The ocean has warmed and become more acidic as a result of anthropogenic carbon emissions, but the combined impacts of these changes on the bulk of marine organisms are yet unknown. The marine ecosystem experiences varying pH and temperature changes. In comparison to lower latitudes, higher latitudes are seeing both more abrupt pH drops and faster temperature increases. Particularly the Gulf of Maine and the Northwest Atlantic are quickly warming. Studies on species in the Northwest Atlantic that have been affected by ocean acidification (OA) are rare in number and even fewer have taken into account pertinent co-stressors such as rising temperatures (Salisbury & Jönsson, 2018). The effects of elevated partial pressures of carbon dioxide (pCO_2) have only recently been explored, despite the fact that temperature is a well-known factor in the growth, development and survival of crustaceans. It is now evident that these reactions depend on the species and life stage. For instance, Dungeness crab (*Cancer magister*) larvae and juvenile Tanner crab (*Chionoecetes bairdi*) larvae both have lower survival rates in high pCO_2 settings, but other crustacean species exhibit no change (Lin et al., 2020). The one larval study discovered that high pCO_2 caused planktonic larvae to develop more slowly and have shorter carapace lengths. Recent research on the larvae of the European lobster (*Homarus gammarus*) has yielded conflicting results. High pCO_2 caused carapace abnormalities in one study, while it had no discernible impact on wet body mass or carapace length in another. Only exceptionally high pCO_2 (9000 ppm) exposures have had a negative impact on juvenile benthic *H. gammarus*.

No studies have looked at the potential interactions between increased pCO_2 and temperature on any life stage, despite the fact that the effects of temperature on the larval growth and development of the American lobster are well understood. These studies were created to close a knowledge gap about how anticipated rises in temperature and pCO_2 may affect the survival, physiology and development of the larval stages of *H. americanus*. The American lobster has a long history of serving as a model organism for crustacean physiology and ecology in addition to being significant from a cultural and economic standpoint. It is an ideal model organism to study the effects of various pCO_2 and warming scenarios on crustacean development and physiology because this research legacy offers valuable background information on fundamental biology, physiology and most recently genetic controls unavailable for most marine taxa. Researchers reared freshly hatched lobster larvae in environments that simulated current and anticipated temperatures and CO_2 concentrations for the Gulf of Maine by the end of the century under a moderate CO_2 emissions scenario predicted by the Intergovernmental Panel on Climate Change (Cetina-Heredia et al., 2015).

Climate change will have an impact on marine resources all over the world, which will have an impact on how fisheries of commercial importance are managed. Understanding how climate change is likely to disturb fisheries biology at each of the following levels is necessary: (a) individual (reproductive potential, larval settlement, spatial distribution); (b) population (carrying capacity, productivity, spatial distribution); (c) multi-species (replacing one fishery with another); and (d) ecosystem (dependent predator species, shifts in community composition). In order to resolve these matters, it is crucial to combine biological, economic and social research components with data from a variety of stakeholders and resource-related variables. As a practical first step in examining the effects of climate change on a fishery, we can highlight the possibilities of this mix of quantitative and qualitative methodologies and summarize management implications. Our findings imply that there can be both favorable and unfavorable effects. Our integrated technique is a step toward quantifying the socioeconomic effects on fishers, their communities and national economies as well as connecting the relationships between various variables and fishing productivity.

CHAPTER 7

Economics of Lobster Culture

CONTENTS

7.1 INTRODUCTION

The goal of management plans and policies in lobster fisheries is typically to maximize the economic gain from harvests, which is best supported by gathering economic data to assess management choices. The sustainable economic yield, which is the long-term, sustainable revenue from harvests minus the expenses of harvesting, is used to quantify the economic benefit from various lobster harvests across the world. Maximum economic yield (MEY) is increasingly being thought of as a formal aim for lobster fisheries, notably those for *Panulirus cygnus*, *P. ornatus* and *Jasus edwardsii* in Australia and New Zealand. In order to assess rules like as catch limits, season length, gear limits and gear type, bioeconomic models that incorporate stock, cost and price information are currently being utilized in lobster fisheries that include *P. interruptus*, *P. argus*, *P. cygnus*, *J. edwardsii* and *Homarus americanus*. The development of management systems used to limit catch has also been influenced by economic theory, particularly through the greater application of market and rights-based systems. These seek to offer rewards for prudent stock management that safeguards future harvests as well as incentives and methods for the transfer of catch to more effective operators. Resource sharing difficulties in lobster fisheries can be resolved using economic methods, with the majority of studies focusing on recreational and commercial interactions (Nguyen, 2009).

Because it is essential to guarantee sustainable harvests, the management of lobster fisheries has historically placed a strong emphasis on biological results. To improve benefits from fisheries rather than just ensure sustainable harvests, management increasingly employs economic approaches, methods and objectives. Although there are many different management strategies used in lobster fisheries, the majority include input control elements like gear and vessel restrictions. Individual output controls are now widely implemented in fisheries, with economic theory serving as the main driving force behind this transformation. This strategy, which has been well-developed in some lobster fisheries in the Pacific and Atlantic oceans, aims to encourage greater resource management by allocating access rights. The use of reference points that aim for maximum economic yield in addition to biologic limit reference points has occasionally allowed lobster fisheries to explicitly incorporate economics into their management framework. The study of fisheries economics helps to address questions of resource allocation and sharing among diverse users. Along with the recreational and commercial harvest sectors, these various user groups also include non-extractive users, who are increasingly catered to through leisure activities within marine protected areas (MPAs). A number of the complicated and new issues that many lobster fisheries are currently confronting can also be resolved with the help of economics. These include controlling harvests in the face of increasing climate change-related uncertainty and taking into account intricate geographical challenges in many fisheries. These examples are meant to show the similarities and differences between various management strategies rather than to be comprehensive case studies or literature reviews on any one fishery or even to cover the most profitable fisheries in the world.

7.2 BENEFITS

Net economic benefit, which is simply the overall gain from fisheries less the expenses of harvesting, including opportunity costs like unpaid labor, gives a formal assessment of the economic output of fisheries. The main objective of the study and management of these fisheries is to provide benefits to commercial fishermen and communities who depend on fishing. The objectives of fisheries management legislation nearly always include this intention to deliver economic benefits, which has an impact on the goals of lobster fishery management plans. Despite the fact that this goal is widely shared, there are many different ways that management tries to create economic gain or how economic success is monitored and evaluated. All of these layers must be properly constructed for a commercial fishery to be economically effective, which is a difficult task for any fisheries management system.

It necessitates taking into account the economic implications of the entire set of rules and regulations that create the operating environment for the fishery, including the management of harvest size, the regulations governing size restrictions and seasons, as well as the size and number of operators. Bioeconomic models, which combine resource biology and ecology with the economics of fisher behavior, can be used to take these potential consequences into account. The variables of time, space and uncertainty can all be included in bioeconomic models. Not only do lobster populations benefit commercial fisheries, but also native fisheries. Because it offers a mechanism for making decisions in regard to non-market values, economics has importance in the management of such fisheries. To identify the best distribution of catch between the commercial and recreational fishing sectors, for instance, one might measure the usefulness of recreational fishing. Setting goals that may be used to gauge progress toward objectives and existing conditions is essential for effective fisheries management. These goals can be related to a wide range of fishery-related topics, including stock measurements, ecological assessments, social consequences and economics. Even though economic benefits are significant, specific economic aims for managing lobster fisheries are rare. The Torres Strait tropical lobster *Panulirus ornatus* fishery is one example from Australian Commonwealth Fisheries, although the majority of other Australian lobster fisheries are controlled individually under the purview of individual states (Plaganyi et al., 2018). The goal of the fisheries harvest systems in this jurisdiction is to maximize commercial yield at levels of exploitable biomass. The harvest control rule for the Torres Strait *P. ornatus* fishery consists of a constant exploitation rate, which aims to approximate the level that yields MEY. Since it is the yield that can be obtained during each period of time that maximizes the long-term net economic benefits of the fishery, MEY is the theoretically ideal sustainable yield i.e., the quantity of lobsters that can be harvested in a particular period, typically a season or a year. This strategy is also being used to make improvements to the harvest strategy for the *Panulirus interruptus* fisheries off the coast of Baja California, Mexico and the *Panulirus cygnus* fishery in Western Australia.

7.3 ECONOMIC ANALYSIS

Economic analysis has been utilized in other lobster fisheries to evaluate management plans that weren't directly aimed at MEY but still had this overall goal in mind. The Tasmanian *Jasus edwardsii* fishery has set a target for its catch per unit effort (CPUE) by 2020, which is roughly equivalent to a stock rebuilding pathway to MEY. Many *J. edwardsii* fisheries in New Zealand have harvest plans with a target CPUE (Breen, 2009). The industry had competing goals of high capture and quick stock rebuilding in two of these fisheries; in this situation, the effects on economic yield helped assess various solutions. Typically, capture rate, biomass, or egg production are used as performance indicators to compare results to goals in lobster fisheries. Beyond just the change in gross product value, the economic performance of the lobster fisheries is infrequently reported in accordance with economic targets. The fisheries for both southern rock lobsters *J. edwardsii* off Tasmania, Australia and eastern rock lobster *Sagmariasus verreauxi* off New South Wales (NSW), Australia, are rare cases where economic performance is monitored. Both are controlled under individual transferable quota (ITQ) systems. The total annual catch allowed in these fisheries is divided into smaller parts and percentage shares are exchanged on a market. Thus, market transactions show the value that quota owners place on these units, which is then used as a stand-in for economic yield in both of these fisheries. Because the market value of the shares might change depending on variables unrelated to the lobster stock, such as the availability of business loans and market functioning, it is merely a proxy for economic yield and not a direct indicator. In particular, the decrease in catch caused the stock to rebuild, which boosted capture rates and decreased the effort needed to take the catch. Because less effort translates into cheaper fishing costs, this increased the market price of the quota units.

The two *J. edwardsii* areas in South Australia may provide the most thorough reporting of economic metrics in a lobster fishery. These evaluations cover changes in employment, contribution to the GDP and economic output. In these lobster fisheries, the economic output of the fishery is assessed using a survey procedure in which expenses and earnings are directly received from the fleet. When compared to typical performance metrics like aggregate catch or catch rate i.e., the average catch per unit effort, this can offer a new perspective on fishery performance. There was tremendous impetus for management reform in the Northern Zone fisheries since the fleet's economic yield has been negative for the past nine years. The management's solution was to reduce total capture so that stocks and catch rate would grow and the cost of harvesting would decrease. Managing economic yield involves dealing with uncertainty and change in the fisheries as well as the aim as well as the route to get there. A process known as management strategy evaluation (MSE), which involves evaluating the outcomes of a variety of management strategies to uncover performance trade-offs, can be used to investigate the management decision pathway. MSE models are a component of a simulation-based adaptive management cycle.

Instead of attempting to dictate the best course of action, MSE application aims to enlighten the decision-maker about the repercussions of each alternative. Critical decision points and significant assumptions are usually determined using MSE during a retrospective analysis. The biological and economical goals of the *J. edwardsii* fishery in New Zealand have been accomplished through the application of this approach. MSE has not been widely used in lobster fisheries elsewhere and it is distinct from but complementary to bioeconomic modeling, suggesting that increased collaboration between economists and fishery scientists is likely to be the most helpful strategy for influencing management decisions. Biological and fleet data are typically combined with cost and revenue data when economic modeling lobster fisheries. Although the relationship between tonnage harvested and revenue is straightforward, two characteristics of the relationship between stock and economic results are crucial for comprehending the relationship between catch and economic yield. The first is known as stock externality, which refers to the decrease in harvesting costs when the stock is more plentiful and, consequently, catch rates are higher. The second is known as the marginal stock impact and it refers to the reduction in sustainable output and consequently income that might happen as a result of growth and recruitment overfishing. Numerous characteristics of lobster fisheries can be modeled to look for management improvement opportunities. The size of the individuals taken can be changed to manage the economic yield per recruit because individual lobsters are simple to quantify and release with low discard mortality. Price can fluctuate depending on size, season and physical characteristics like color, allowing for economic gains to be made by varying the period of harvest.

Economic analysis of the cost element of lobster fisheries is also a good fit, particularly for fisheries based on traps when trap standards are in place. Costs can be easily calculated by surveys and scaled with effort when the effort is measured in standardized effort units. Through trap management, there appear to be significant chances to raise economic yield in many lobster fisheries. More effective trap use might cut the cost of harvesting in a *Homarus americanus* fishery by 25% (Acheson, 2001). The Maine *H. americanus* fishery has successfully implemented trap constraints, but doing so was not an easy task. Initial attempts to limit the total number of traps actually resulted in an increase in trap sets and, consequently, fishing costs because of unexpected social responses, a common issue in input-controlled fisheries. Because revenue and cost vary by gear type, modeling of fisheries with several gear types is more difficult. This problem was significant in a bioeconomic analysis of the Panulirus argus fishery on the Yucatan Peninsula shelf, where gillnets, artificial habitats, scuba and hookah diving and traps were used during the adult migratory. Profits from the use of diverse fishing techniques and equipment, which were connected to various selectivity and costs, were reported to differ significantly. In this fishery, the proportion of catch taken by various gear types has changed, with a greater proportion now harvested from casitas. However, the challenge of incorporating various gear types into models for bioeconomic development in the area makes this process more difficult. With the assumption of an exponential catch function, multiple studies in Florida have approximated the MEY for *P. argus* using the surplus production method. This functional form demonstrated that greater levels of effort do not increase overall catch and matched the observed historic catch and effort data. The bioeconomic MEY solution suggested a significant decrease in the effort of at least 50% would be required to maximize resource rents i.e., the share of profits that are assumed to be generated by the value of the natural stock when this information was combined with the estimated fixed cost and cost per unit of effort from data obtained through surveys. Dynamic age-structured bioeconomic models have been used to predict the impact of various management strategies on the spiny lobster fisheries of Brazil and Cuba in the West Central Atlantic (Rafael, Vázquez, Martinez, & de León, 2005). The use of reference points with bioeconomic model input was recently included in management policy for the *P. interruptus* fishery in Baja California, Mexico.

Open access will continue to draw new users as long as these users continue to receive positive private net advantages, just like with any common property resource. The majority of fisheries around the world have developed in this manner; as economic theory predicts, profitable fisheries draw new participants and fisheries are most profitable when there are few harvesters catching a lot of fish. Fisheries for lobster are not an exception. Fishing effort has increased in profitable fisheries when it comes to commercial lobster fisheries. The number of participants expanded significantly during 15 to 20 years in the late twentieth century in both of the main U.S. commercial lobster fisheries, American lobster *H. americanus* and Caribbean spiny lobster *Panulirus argus*. Up until the mid-1990s, the rise in participants was accompanied by a growth in yearly catches of American lobster, although annual landings of Caribbean spiny lobster remained largely stable. Despite variations in participant numbers, the fishery's vessel-level conditions remained the same: too many boats were pursuing too few lobsters. Fisheries managers believed that restricting access to the fisheries would stop this unfavorable trend. At its most basic level, limited entry merely forbids the addition of new users and is often put into practice at the vessel level. As the simplest management method for limiting the number of participants, restrictions on the total number of vessels allowed to operate are used in almost every lobster fishery today. The total number of licenses is often regulated, but since they are transferable, new people may be permitted to join the fishery as existing ones are closed. Although many lobster fisheries have imposed entry restrictions, reducing just one aspect of fishing effort such as the number of boats, permits, people, etc. rarely results in a reduction in the overall catch. This is due to the diversity of factors involved in fishing that affect overall catch and vessel level efficiency. The size of the vessel, the fishing site, the time of year, the gear and the length of the trip are among the variables that affect the catch at the trip level apart from the weather, ecological changes and macroeconomic situations. Due to a variety of variables, fishermen can typically expand in other regions to make up for limitations in any one part of their fishing activities.

7.4 RESTRICTIONS

Fisheries managed to restrict admission but otherwise leave fishing effort unconstrained are referred to as controlled open access fisheries. If the fishing effort is not adequately constrained, fisheries operating under limited entry may perform no better than open access fisheries. Participants in such fisheries keep up the competition for the resource by putting more effort into other areas of fishing. This increased use of different fishing operations strategies as operators compete includes ineffective additional equipment investment. If other components of fishing effort are not also handled in lobster fisheries, limiting the number of vessels or individuals should not be expected to increase the profitability of the fishery. Four components make up effort restriction or input control systems (although each of these elements is not always applied). These include (i) limiting the quantity and type of gear used, (ii) limiting the time and area in which this gear can be used, (iii) managing the increase in effort effectiveness over time as fishing techniques improve, such as through technological advancements, and (iv) managing the transferability of effort between operators. Theoretically, successful use of these components can enable managers to shift the fishery from one that is operating at open access to a level of effort that resembles maximum economic yield. Economic studies of the lobster fishery regularly show that reducing fishing effort, especially with effort limitation regimes, has a significant positive economic impact.

In effort-limited lobster fisheries, many aspects of trap design can be regulated to increase efficiency. These include the maximum dimension, the diameter of the opening, restrictions on the quantity of juvenile lobsters used as attractants, the distance between slats and the use of escape gaps or rings. These rules not only limit effort but also allow bycatch and young lobsters to escape. Regulating gear, however, frequently has unfavorable economic effects since it drives up fishing expenses by lowering capture rates and raising the price of building traps. It prevents trap design innovation, which would otherwise lower fishing expenses. A limit on the amount of gear available, which is applicable to both trap and other methods of lobster capture, such as trammel nets, is a common effort control measure in many lobster fisheries. These fisheries include those for *Palinuris elephas* in the Mediterranean Sea and *Panulirus homarus* in the western Indian Ocean. The design, quantity and frequency with which the equipment is deployed throughout the year all affect the fleet's potential to harvest. As a result, seasonal and geographical harvest constraints are frequently added to effort limitation systems. Because restricted seasons can induce pulses in supply and consequently lower prices, economic efficiency is increased where limitations are put on the total amount of gear accessible to the fleet rather than the number of days fished. Closed seasons and vessel restrictions also suggest that capital invested in vessels and processing facilities is not being utilized to its maximum potential, which is inefficient but may be justified to protect spawners and or juveniles.

In addition to gear design, quantity and temporal constraints, there are many other elements that influence production and harvesting capacity. These factors in lobster fisheries include, among others, the skill and experience of the fishermen, technological advancements such as the switch from synthetic rope to electronics like GPS, the geographic range covered by the stock, problems with enforcement and the rigor with which the traps are managed at the vessel level such as the quantity of trap pull. Many of these elements contribute to a phenomenon known as effort creep, in which the efficiency of an effort rises over time. For input-controlled lobster fisheries, effort creep can be a significant issue. Reducing the total number of traps permitted in the fishery, or establishing a total acceptable effort threshold either at the vessel level or fleet level, is one method for reducing active effort and compensating for effort creep. From an economic standpoint, the capacity to exchange or transfer effort between persons in the lobster fisheries would be advantageous since it encourages fleet rationalization since the traps or pots can be transferred to more effective operators, which represents a better use of capital. Transferability enables the development of a market for labor such that the value of the traps is correlated with potential catch flows. One of the elements of output or quota systems that is advocated is the relationship between fishers' wealth in fishing assets and the future health of the fishery since it rewards resource stewardship.

There are numerous instances of individual transferable effort (ITE) schemes or tradable effort limited lobster fisheries globally. One early example is the Florida lobster fishery for *P. argus*, when such a system was implemented for economic rather than biological reasons to increase fisheries productivity. This was necessary since catch rates dropped and individual operators' income decreased as a result of the almost 10-fold rise in trap numbers caused by free access from the middle of the 1970s to the beginning of the 1990s. The first steps in addressing the issue of diminishing catch rates and a related rise in average cost were capping the number of traps, which was followed by a reduction in trap numbers to avoid crowding of gear. By mandating that each trap be marked with a different number, it was possible to cap the number of traps. Following capping, a procedure known as active reductions was used to lower each individual's holdings by 10% over the course of three seasons. As a result, the fishery had fewer traps, but the method was unpopular and was short-lived. In 1998, passive reductions, in which 10% of traps were taken out when operators changed, were combined with the return of active reductions in pot numbers at a lower level of 4%. Although switching from active reductions to a passive program is favored by the industry, it is ineffective at reducing effort. The number of traps had decreased by about 40% by 2011; passive reductions are expected to continue until the number of traps decreases by another 20% (Ritchie et al., 2013).

7.5 MANAGEMENT

The management strategy utilized in Florida was largely replicated in 2003 with *H. americanus* in a region of the Northeastern United States and in 2011 with the switch from ITE to output management in the *Panulirus cygnus* fishery in Western Australia. A predetermined and constrained number of units were allotted to fishermen under the Western Australian ITE system. The number of these units, which originally corresponded to the number of traps, represented the right to utilize a trap. Later, the units and traps were split such that changes in the ratio of fishery access units to physically deployable traps could be made to alter the overall number of traps in the fishery. Transferring equipment between operators, allowed for a decrease in the overall number of traps used in the fishery and promoted fleet rationalization. Although the *P. cygnus* ITE system had the potential to boost economic production from the fishery, a number of application-related issues led to the system's discontinuation in 2011 in favor of ITQs. It was politically challenging to set the trap: unit ratio at the ideal level for economic yield in the Florida fishery since ITE system reductions in trap limits were controversial. According to a bioeconomic analysis of the fishery, a significant drop in the effort of almost 40% from 2007–2008 levels was necessary to maximize economic production. Even if this effort shift had been implemented, the requirement to address effort creep in ITE systems meant that opposition to effort reduction would still be a persistent issue. As fishers continued to compete with one another for catches, effort creep was also a sign of inefficient capital allocation.

With peak landings occurring around times of greatest supply availability, these examples of ITE systems responded to stock availability rather than market demand. This covered the start of the season, catchability peaks and stronger recruitment. With little motivation to maximize the value of the catch, this supply-driven pattern in fishing produced scarcities in the supply. When it came to the *P. cygnus* fishery, a significant amount of the catch was sold into local markets at lesser prices as opposed to high-end live export markets in Asia. Due to the substantial swings in catch and revenue between years brought on by the increase in catch in years of strong recruitment, firms faced challenging circumstances during following recruitment troughs. Fishermen reduced capture by about 40% at the beginning of the season when the price was low and also targeted better value size classes as a result of the termination of the ITE system in the *P. cygnus* fishery and the transition to ITQ management in 2011. The Florida ITE system also serves as an example of an issue encountered by many lobster fisheries when trying to use transferrable rights systems to help harvest move to more effective operators. Unlike a regulated online market where past transactions are public, this transfer market had an informal market structure, so interested buyers and sellers had to actively seek one another out. Trading in the Florida fishery was impacted by this market's underdeveloped and informal nature, which was exacerbated by its members' varied cultural and geographic backgrounds. The operation of this market has also been hampered by the lack of knowledge of asset valuations and the necessity to spend money looking for trading partners who have equivalent units to trade. A market for trading *J. edwardsii* harvest rights became more and more dominated over time. This leads to information asymmetry and exacerbates the issue of an informal market structure.

The application of size limitations, which are particularly effective due to the ability to handle, measure and release lobsters uninjured, is a significant management method utilized by the majority of lobster fisheries. Size restrictions can be utilized to manage economic yield even if they are typically set for biological reasons. Commercial harvests frequently came first in lobster fisheries before recreational harvests. In these situations, the sequential development of sectors has led to a management structure that does not simultaneously take into account all consumers. Recreational fishers may feel entitled to access as citizen owners of a natural resource that benefits everyone and this de facto harvest right might lead to issues when recreational fishing effort increases. Therefore, in order to maximize the resource's economic value or try to operate at maximum yield, a mechanism for limiting overall effort and redistributing resources among sectors is needed. The administrative approach and the market approach are the two main different strategies for creating inter-sectoral reallocation mechanisms. Any reallocation system must have an initial allocation to each sector and a predetermined unit for transfers across sectors in order to function effectively. For user groups using the same metrics, allocations are quite simple and can be thought of from either the input or output side i.e., number of traps and weight of harvest. The fact that gear and management regulations sometimes vary between groups complicates many lobster fisheries when taking into account all user groups. In such circumstances, the marginal analysis necessary to conceptually and economically optimize allocations is challenging.

Managers intervene in the administrative approach to sector reallocation in order to get a more ideal result. Market-based share trading between industries is one such. The case for using market mechanisms is predicated on the idea that resource allocation through a government procedure leads to less efficient economic use of the resource. Promoting the joint business and leisure sectors is one market-based option. Some commercial *P. argus* fishermen in Florida have made an effort to take advantage of the demand from tourists by allowing consumers to board their boats as a tourist attraction. The fishermen paid a greater price for the same lobster while the tourists received fresh lobster that they assisted in harvesting. In the more integrated market structure for recreational allocation recreational management organizations trade in the catch share market on behalf of recreational fishers. Since measuring the marginal benefit of recreational fishing is more difficult than for commercial fishermen, most economic analyses of recreational lobster fishing place more attention on the cost factors. The benefit of commercial fishing is easily measured by the money made, whereas the benefit of recreational fisheries is defined as utility, which includes enjoyment. The benefit of recreational fishermen has the characteristic of diminishing marginal utility, where the first lobster provides the greatest value or delight and this advantage decreases with each additional lobster over the course of a particular period of time. Additionally, the recreational value is frequently not proportional to size, with larger lobsters being valued higher than their weight would imply. The establishment of the total recreational catch can be treated in the same way as total allowed catch limits in the commercial sector, despite the fact that the majority of economic study and debate on recreational catch deals with distribution between sectors. To maximize the net benefit from recreational fishing, commercial fisheries must balance increasing catches against decreasing catch rates. Although it is not explicitly used in lobster fisheries that are fished for recreational purposes, it is theoretically possible to achieve maximum sustainable utility. By permitting more people to participate in the recreational sector while harvesting fewer lobsters per participant, managers try to increase the utility of the sector. In the recreational sector, like in the commercial sector, management of high catch rates is vital because unsuccessful leisure fishing trips have little practical value. However, managers should be cautious of incentives to boost recreational values that, if not intended, also work to reallocate harvest away from the commercial. For instance, managers of the Florida fishery have set up a two-day recreational sport season that begins just before the start of the commercial season; any increases in recreational sector harvests come at the expense of lower commercial harvests in the season that follows. The economic incentives for commercial fishermen to support catch limitations on their industry are diminished by this reallocation.

7.6 FISHING

Non-commercial fishing is done for both recreational and subsistence reasons, i.e., to get food. The Pacific is home to subsistence lobster fishing and in certain cases, the promotion of larger-scale commercial harvesting has been resisted on the grounds that it should be protected. The political challenge of allocating resources between sectors in a way that is economically beneficial is exacerbated by disparate harvesting objectives. Although there are significant supplementary allocations for subsistence harvesting, the *J. lalandii* (rock lobster) fisheries in South Africa are quota regulated. Due to the value of rock lobsters, the line separating commercial and subsistence fishing became increasingly blurred, which had a significant impact on resource management. In many jurisdictions, the distribution of catch among sectors takes into account traditional harvests. Although the catch is typically too small to have an impact on talks about resource sharing or TAC setting, the management of each Australian lobster fishery takes into account the customary catch. The value that the community placed on the rehabilitation of dwindling lobster stocks has been used in Norway to quantify the non-extractive value of marine animals. The benefit or enjoyment that recreational user groups derive from viewing lobsters has been quantified for dive charter tourists viewing *Panulirus argus* in the Turks and Caicos Islands. This benefit is a direct, non-extractive benefit that exists from lobsters. The distribution of resources to various groups is frequently used as a defense for MPAs' obligation to offer an opportunity for people to observe natural concentrations of exploited species. It is crucial for management to account for this stock allocation to MPAs by limiting capture by harvest sectors since displaced catch would otherwise raise harvest rates in open areas, depleting the stock and causing long-term economic loss. However, if the fishery is managed to MEY targets, in which case a net loss must occur, this is not possible.

Commonly, nations and different legal systems share lobster stocks. This most frequently occurs through interactions between larval sources and sinks, which indicate the necessity for a coordinated strategy to govern egg production. In many lobster fisheries, such as the European fisheries for *Palinurus elephas* or the western central Atlantic fisheries for *Panulirus argus* with source-sink dynamics and harvesting taking place across Cuba, Brazil, the Bahamas, Honduras, the USA, Nicaragua, Mexico and Belize, issues of dividing catches between jurisdictions are complex (N. Ehrhardt, 2005). The targeting of economic yield in the management of the West Central Atlantic spiny lobster fisheries is complicated by four spatial factors: (i) it is a transboundary resource; (ii) it is characterized by a metapopulation with source-sink configurations; (iii) gears and fishing methods are heterogeneous with varying effects on population structure; and (iv) the heterogeneity in management strategies, regulations including different legal sizes and trade restrictions and (v) the heterogeny. When lobsters of harvestable size travel across borders, as happens in the Torres Strait tropical rock lobster *Panulirus ornatus* fishery during yearly migrations, resource sharing becomes increasingly difficult and controversial. Additional allocation problems within this fishery exist between the trawl sectors as well as between indigenous and non-indigenous harvesting, which may necessitate the harmonization of bilateral agreements. To help with the resolution of these intricate problems, current research has started to look at the economic ramifications of the numerous allocation possibilities.

Even though these restrictions might diminish efficiency and result in lost economic return, many resource sectors are regulated within restraints meant to achieve social goals. Frequently, the goal is to advance certain facets of local development, such as jobs and community networks. Examples include the promotion of local labor in the mining sector over "fly-in, fly-out" labor despite higher costs or the distribution of water allocations among irrigation regions to share benefits even when the result is less effective than the concentration of irrigation schemes. In addition to economic and environmental considerations, management strategies for the lobster fishery frequently mention social outcomes. There have been few formal attempts to impose legal social constraints around the management of lobster fisheries, despite the fact that the demand for social results from lobster fisheries is widespread at the governance level. Individual ownership restrictions are occasionally used in lobster fisheries to avoid a concentration of ownership. This not only contributes to the apparent societal advantage of having ownership distributed among many people, but it can also help reduce the asymmetry of market power. This social advantage involves sustaining coastal fishing villages and retaining the small-scale nature of the fishery. By allowing exceptions to transfer fines and costs to family members, these management objectives for the Florida *P. argus* fishery have been expanded. Although well-intentioned, these ownership restrictions can be challenging to uphold. Furthermore, if there are strong familial ties within a group, concentration control can be difficult.

One argument against aiming for maximum economic yield with ITQs is that rents from these communal assets can easily leave the area, as in the case of catch share owners who reside abroad. Since interstate pension funds have bought ITQ catch shares in *J. edwardsi*i fisheries, there has been a boycott of leasing markets by active fishermen. The South Australian lobster fisheries economic assessments estimate state and regional economic impact using input-out analysis. The fate of economic rents from fisheries is crucial to this discussion because, while rents that flow out of the community to distant owners of the catch shares can still have a positive impact on the community, doing so will make that impact less noticeable. When there are significant interactions with other species, such as when bycatch from lobster fishing adds to the value of the fishing operation, or when fishers use capital across multiple fisheries, such as lobster, scale fish and scallop, the efficient use of any stock is complicated. The effects of shared inputs such as equipment and vessels and outputs on the performance of fisheries can be best determined using bioeconomic models such as where bycatch is dependent on the targeting of lobsters. Bioeconomic techniques can assess other benefits, such as ecological services and recreation, in addition to the goal of an effective commercial harvest. Predator and prey species are taken into account while assessing the potential effects of fishing on ecosystems.

The majority of predator-prey species bioeconomic research conducted to date has focused on situations in which both species are the targets of commercial endeavors. An integrated bioeconomic model is necessary in order to identify an effective management strategy for these linked fisheries since economic or environmental changes in one can have an impact on outcomes in the other. There are also several excellent studies that might be applied to lobster fisheries that use bioeconomic models to investigate the ecological relationships related to managing fisheries. Models like Atlantis and Ecosim can be used to examine wider ecosystem interactions of lobster fisheries. To get insight into the interactions of species within the habitat, this strategy was used using *P. polyphagus* and other Panulirus species trawled off the Southeast Indian coast. Even though these models were created to study interactions between ecosystems, more recent advancements have added economic components. Marine Protected Areas (MPAs) (Figure 7.1) are used in coastal management for a variety of goals, including improved diving chances for leisure and as research locations for keeping track of unfinished populations. The resource distribution between non-extractive and extractive uses is conceptually analogous to the conflicts between recreational and commercial fishing since these non-extractive uses have economic value.

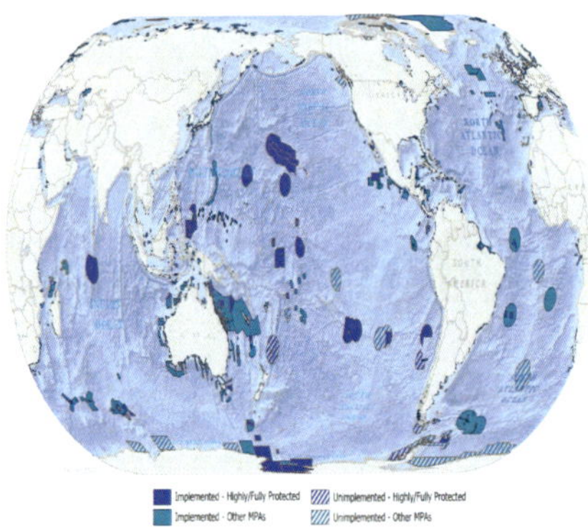

Figure 7.1. Marine Protected Areas.

Source:https://upload.wikimedia.org/wikipedia/commons/thumb/4/4f/GlobalMPAs_logo_2020.jpg/1200px-GlobalMPAs_logo_2020.jpg

Displaced catch from MPAs typically leads to reduced catch and lower revenue and lower catch rates (greater costs), which suggests a loss of economic yield. Due to the fact that the relocated catch is not moved elsewhere, as demonstrated by field investigations of *Palinurus elephas*, this typical consequence does not occur when the lobster stock is severely overfished. Modeling of the impact of MPAs on quota-regulated *Jasus edwardsii* fisheries is in line with the general model and demonstrates that displaced catch from MPAs decreases overall biomass and egg production unless there was a proportional reduction in the allowable catch. Even when the catch is stable, MPAs can nevertheless result in economic loss, as seen by the necessity for longer travel distances to fishing sites that are still accessible for *P. argus* fishing in Nicaragua. The economic impact of MPAs on lobster fisheries, whether through an increase in fishing costs or a drop in revenue from quota reductions, has an impact on rights-based management. This is due to the weakening of the exclusivity and security characteristics FAO considers crucial for these systems because of the decreased economic output. The siting of MPAs can have a big impact on stock results since lobster fisheries are often regionally varied in terms of biology and fleet. Future bioeconomic modeling and analysis of alternate area closure strategies for managing lobster metapopulations with established source-sink dynamics might take this strategy into account.

Due to the inter-annual variability in productivity, lobster fisheries typically require adaptive harvesting and management. This strategy will become more crucial in regulating the effects of climate change on lobster populations, including modifications to the productivity and distribution, such as the recent fall in the recruitment of *P. cygnus* and *J. edwardsii*. Numerous factors, including shifts in ecosystem productivity, growth, recruitment and reproduction, can have an impact on lobster fishery's production. Effects through ecological connections are inherently complicated; for instance, the climate change-related rise in J. *lalandii* abundance off the South African west coast fishery is reducing abalone harvests through interaction with urchins. Even for single species, the effects of climate change are difficult to predict because biological and ecological processes can work against them. For example, the *Panulirus penicillatus* fishery off Taiwan's eastern coast is expected to experience increased growth, decreased recruitment and increased mortality.

7.7 MODELING

Bioeconomic modeling, which is often based on historical data, may face additional hurdles as a result of climate change uncertainty. It has also been demonstrated to decrease fishermen's capacity to make informed choices in a co-managed *J. edwardsii* fishery. In the best-case scenario, data to evaluate the consequences of historical climatic changes only span a few decades, therefore they are not a suitable benchmark for predicting future outcomes. Quantitative techniques, such as Monte Carlo analysis, can be used to calculate the odds of exceeding Limit Reference Points (LRPs) with various management techniques. They are helpful in estimating the likelihood that economic LRPs for performance indicators may be exceeded as a result of evolving economic and/or environmental uncertainty. The use of decision theory for systematic choice under uncertainty taking into account various risk attitude criteria, with and without mathematical probabilities, is another alternative for recognizing the associated uncertainties in the decision-making process of managing lobster fisheries.

CHAPTER 8

Pollution and Lobster Aquaculture

CONTENTS

8.1 INTRODUCTION

Atomic absorption spectrophotometry (Figure 8.1) was used to calculate the amounts of heavy metals. Statistically significant variations between study concentrations in sediments during pre and post-monsoon at sample locations were found.

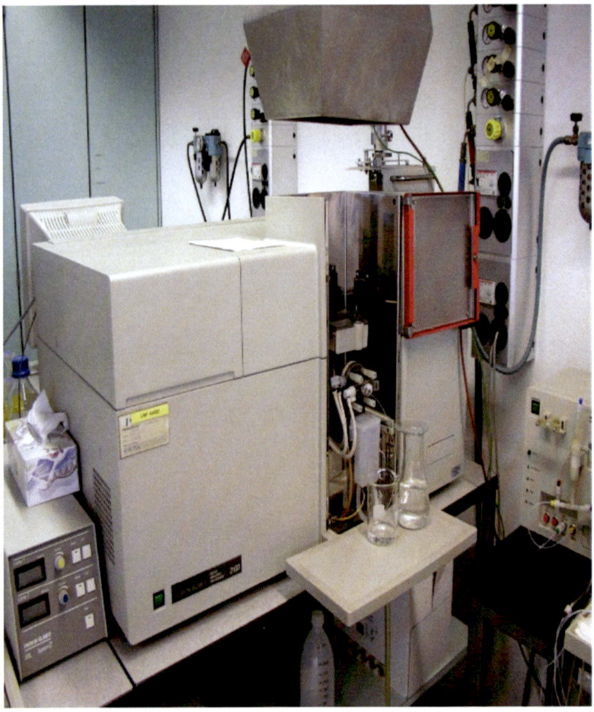

Figure 8.1. Atomic absorption spectrophotometry.

Source:https://upload.wikimedia.org/wikipedia/commons/thumb/5/51/FlammenAAS.jpg/800px-FlammenAAS.jpg

With the exception of copper in the muscle of *P. homarus* in the post-monsoon, analysis of variance showed a significant difference in *C. arel* and *P. homarus* across the different sites in the pre- and post-monsoon. The tendency in the mean metal concentrations was Cu> Pb> Cd> Hg, Cu> Pb> Cd> Hg and Cu> Hg=Pb> Cd, respectively, in the sediments and muscle of *C. arel* and P. homarus. Additionally, the order of the samples' mean metal concentrations was sediment > *P. homarus* > *C. arel* (Mahdi Abkener, 2020). But only cadmium in the muscle of *P. homarus* exhibited a significant difference in the paired t-test analysis comparing heavy metal concentration in pre- and post-monsoon. Overall, a comparison of the study's findings showed that the Gulf of Chabahar and Jask Port had higher concentrations of lead, copper, cadmium and mercury in sediments and muscles of *C. arel* and *P. homarus* than other parts of the Iranian beaches around the Oman Sea that were also investigated.

With several unidentified or poorly understood effects on species and ecosystems spanning land and aquatic settings, light pollution is a fast-expanding risk to biodiversity. Despite the quantity and importance of invertebrates to marine and freshwater ecosystems, the impacts of artificial light at night on aquatic invertebrates are poorly studied when compared to other taxa. To test the impact of artificial light at night on this species' physiology and behavior in comparison to appropriate controls, researchers attached heart rate and acceleration biologgers to adult Caribbean spiny lobsters (*Panulirus argus*), an ecologically, culturally and economically significant species in the western Atlantic Ocean. In order to imitate the light conditions offshore of urban centers, the experiment was carried out in a reef mesocosm in the Bahamas using incandescent lights set to 1 lux. Artificial night-time light was shown to have no impact on its heart rate or locomotor activity. At both short-term and long-term temporal periods, we saw a separation between activity and heart rate. Heart rate did not change with the lunar cycle, despite the fact that lobsters were more active at night. Despite the average amount of time spent active nearly tripling from days to nights, to 19.5 percent from 7.2 percent, there was only an 8 percent variation between the average heart rate during the day and night. Results imply that *P. argus* might be somewhat resilient to low levels of light pollution, which justifies future study into several facets of this species' life history, performance and fitness in the face of this possible anthropogenic disturbance.

8.2 LIGHT POLLUTION

Globally, the damage posed by light pollution to terrestrial and marine biodiversity is expanding quickly. The effects of artificial night-time lighting on animals and ecosystems have been extensively studied. These effects include interference with sensory perception, the diel cycle and ecological interactions. Such disturbances are supported by physiological mechanisms, whose investigation might provide insight into how organisms react to light pollution, especially when its consequences are not immediately apparent to the naked eye. However, light pollution (Figure 8.2) poses a threat to marine and freshwater species and is less well understood in these aquatic ecosystems.

Figure 8.2. Light pollution.

Source:https://scx2.b-cdn.net/gfx/news/hires/2015/coastallight.jpg

Figure 8.3. *Perca fluviatilis.*

Source:https://upload.wikimedia.org/wikipedia/commons/thumb/0/09/Perca_fluviatilis2.jpg/800px-Perca_fluviatilis2.jpg

The majority of studies on the topic has been on terrestrial systems. Animal responses to extrinsic stimuli, particularly those with anthropogenic origins such as light pollution, which together add to allostatic stress, are evaluated using a variety of biomarkers. Understanding how anthropogenic disturbances affect functionality, fitness and population-level ecological processes is particularly interesting in the setting of wild species.

Figure 8.4. *Albula vulpes.*

Source:https://upload.wikimedia.org/wikipedia/commons/thumb/8/84/Bonefish_Albula_vulpes.jpg/1024px-Bonefish_Albula_vulpes.jpg

Numerous fish species, including European perch (*Perca fluviatilis*) (Figure 8.3), bonefish (*Albula vulpes*) (Figure 8.4), Atlantic salmon (*Salmo salar*) (Figure 8.5) and baunco (*Girella laevifrons*). Despite their ubiquity and essential role in aquatic ecosystems, the impact of light pollution on aquatic invertebrates has only been the subject of research. The Japanese spiny lobster (*Panulirus japonicus*) has been reported to exhibit activity suppression at extremely low levels of artificial light at night and moderate levels of artificial light at night have been found to alter behavior in four species of North American crayfish. We are aware of only one study to date that examined the hormones in aquatic invertebrates exposed to artificial light at night in an effort to determine if it affected allostatic load and it found that despite altering the behavior and activity levels of two species of North American crayfish, hemolymph serotonin levels were unaffected. Other endpoints, like heart rate (fH) and locomotor activity, can be used to assess stress reactions more constantly and with better resolution. While hormone levels can indicate an animal's status over a certain period of time prior to sampling (Steell et al., 2020).

Figure 8.5. *Salmo salar.*

Source:https://sites.google.com/site/salmonwiki/_/rsrc/1368175217702/salmonidae/salmo/salmo-salar/Atlantic%20Salmon2.jpg

Numerous animal taxa, including decapod crustaceans, have employed heart rate as a proxy for physiological condition and allostatic burden. Changes in fH have been demonstrated to be a valid indicator of physiological conditions in many decapod species. Decapods have the most complicated circulatory system among crustaceans, similar to that of basal vertebrates. Similar to fish and other vertebrates, decapods may experience bradycardia or acardia when subjected to a physiological challenge, depending on the length and severity of the challenge. Historically, only tethered or restrained animals could be used to quantify fH in decapods; however, researchers created a unique method using fH biologgers to measure fH in free-moving decapods. Furthermore, improvements in acceleration biologger technology (i.e., accelerometers) have made it possible to remotely measure activity in small, free-moving animals. Activity is a key factor in determining fH in decapods and a measure of a behavior's response to a stimulus.

8.3 SPINY LOBSTERS

Spiny lobsters are known to live along the southern Javan coast. As a result of oil pollution brought on by oil leaks from ships delivering crude oil to the Cilacap refinery, hydrocarbon accumulation commonly happens in the coastal waters. As an oil shipping route, Cilacap's coastal areas frequently experience oil spills. Sediment in the area with risk levels between low and medium-low can serve as an indicator. Therefore, it was discovered that some areas had a bigger impact on the environment, posing a considerable risk for marine organism survival. The existence of oil poses a threat to the survival of various aquatic organisms, including the spiny lobster. Because the spiny lobster is a financially valuable commodity for generating high-nutrition food, its population needs to be preserved. In light of the circumstances, it is crucial to identify a solution to the issue. The biological impact of excessive coastal pollution on spiny lobster should be the main focus of the investigation. The lobsters were randomly gathered from the waters of the southern districts of Java using gillnets with a 1-inch mesh size. There is a statistically insignificant difference between the numbers of males and females in terms of sex diversity. Despite the fact that the research area's environmental quality was quite poor, ovigerous females were discovered to make up roughly 12% of the population. These data backed up the theory that lobsters have a special adaptation that allows them to live in environments with exceptionally low environmental quality owing to marine oil spills (Haryono, et al., 2018).

Nephrops norvegicus was simultaneously exposed to cadmium, copper and zinc. Exposure levels were set at control, 1, 5 and 25 g/l for cadmium, copper and zinc and 8, 40 and 200 g/l for cadmium (Canli & Stagg, 1996). In homogenates of the gill and the hepatopancreas, concentrations of cadmium, copper, zinc and metallothionein were assessed. Differential pulse polarography was used to quantify metallothionein. In response to increases in exposure concentration, cadmium concentrations considerably rose in the gill and hepatopancreas of both male and female mice. Males, but not females, had much higher concentrations of copper and zinc in their gills. Neither copper nor zinc significantly altered the metal concentrations in the hepatopancreas. Both males and females exposed to metals had significantly higher quantities of the protein metallothionein in their gill and hepatopancreas. Male and female hepatopancreas and gill concentrations of cadmium and metallothionein showed a favorable correlation. Males demonstrated positive associations between MT concentrations and copper levels in the hepatopancreas. The results of this study revealed that, unlike copper and zinc, cadmium MTs in the gill and hepatopancreas of *Nephrops norvegicus* may be utilized as a sensitive technique to detect cadmium pollution in the lobsters. As a function of MPF concentration and food availability, the effects of MPF on the survival, molting and oxygen consumption rates of larval (I-III) and post-larval (IV) stages of the American lobster, *Homarus americanus*, were quantified. Early larval survival was only affected at the highest MPF concentration. Across MPF treatments, MPF had no effect on the timing or pace of molting. Stage II larvae and stage IV post-larvae demonstrated the highest and lowest accumulations of MPF under the cephalothorax carapace, respectively, even though all larval and post-larval stages did so. Stage I larvae only consumed MPF at low MPF concentrations under specific conditions; MPF intake increased with the larval stage and with MPF concentration. Only in later larval stages when exposed to high MPF concentrations were oxygen consumption rates reduced. Overall, our findings show that larval stage, MPF concentration and the presence of food affect MPF interactions and effects on American lobster larvae.

8.4 METALS

On the Pacific coast of Mexico, female and male lobsters, *Panulirus inflatus*, were examined to evaluate the concentration of Fe, Mn, Ni, Cu, Co, Cd, Cr and Zn in their tissues. Different metal concentration discrepancies between tissues and between sizes were detected. In general, viscera had the highest Zn values, whereas gill and gonadal material had the highest Cu concentrations. The exoskeleton, eggs and gills have the highest quantities of Mn, Ni, Cr and Co. Only muscle and viscera contained low but detectable levels of cadmium in the majority of tissues. Only Zn found in the muscle of males and females, Zn found in female gonads, Mn found in male gonads and Cu found connected with male muscle showed a positive connection between size and metal concentration. In recent decades, marine aquaculture has become more significant in the majority of nations. It is vital to consider the environmental effects of this sector in order to develop it sustainably. Marine cage lobster farming has a reputation for producing good returns in Vietnam. With repeated disease outbreaks and elevated lobster mortality, the sector has recently been dealing with sustainability challenges. These occurrences are associated with nutrient contamination, which is brought on by an over reliance on feed supplies. According to reports, the industry could lose up to $30 million per year. Local lobster growers have responded to these problems by spending more money on pesticides and antibiotics or by working harder to keep their cages clean. This behavior indicates that farmers believe there is a conflict between easing environmental pressure and boosting the sector's economic success. This research employs a Material Balance Principle based Data Envelopment Analysis approach employing datasets of marine cage lobster farms in Vietnam to determine the relationship between cost and environmental efficiency. The results demonstrate that increases in input use efficiency would lead to cheaper production costs as well as better environmental outcomes. Given the facts on input prices, it would be advantageous for the environment if lobster farms became more cost-effective and used a more suitable input mix. Similar to this, enhancing environmental performance typically lowers production costs.

Extracellular neurophysiological recording methods were used to analyze *Homarus americanus*. Legs exposed for 3-5 min to drilling mud concentrations of 10 mg/L suspended in seawater changed the responses of 29% of the chemoreceptors tested to food scents; a comparable exposure to drilling mud concentrations of 100 mg/L interfered with 44% of all receptors tested (Derby & Atema, 2011). Despite being identical to one another, the effects of both of these concentrations are statistically significant. A response to interference typically showed a noticeably reduced number of action potentials. The exposure to drilling mud in one preparation altered the temporal pattern of the spikes without changing the overall number of spikes. The 10 mg/L drilling mud itself stimulated more chemosensory neurons. The responses to feeding stimuli were recorded from the legs of lobster that had been exposed to drilling mud for 4–8 days prior to the neurophysiological procedures, indicating that not all chemoreceptors are blocked by these drilling muds. The usual feeding activity of lobster is elicited by leg and antennal chemoreceptors. There is little concrete evidence to support a causal link between chemoreceptor interference and behavioral deficiencies, despite behavioral experiments showing that feeding behavior is altered after exposure to drilling muds and petroleum fractions. The two methods work well together as pollution detection tests, possibly indicating a shared interference mechanism.

8.5 MARKET

Ovigerous and market-sized lobster catches for each haul of traps were kept separate. Regional trends and catch trends in the bay were compared. According to the findings of correlation studies, the feeding/fallow periods on the fish farm had a significant impact on trends in ovigerous capture. Between trends for bay and LFA lobster landings, there was no discernible association. Regardless of the year-to-year variations in water temperatures and weather, patterns of lobster catch per unit effort in Port Mouton Bay appear to be impacted by proximity to the fish farm. Potential variables impacting lobster relocation include odors and habitat changes near open-net pen finfish activities. According to various international organizations, there is a real risk of Hg poisoning from eating seafood, thus pregnant women and young children should avoid or severely limit their consumption of fish and shellfish with high mercury levels (Taylor, Golding, & Emond, 2016).

Figure 8.6. *Panulirus interruptus.*

Source:https://animaldiversity.org/collections/contributors/paul_detwiler/Pinterruptus1/medium.jpg

High mercury levels were found in crustaceans after examinations of two common species, European hake and Norway lobster, that were caught in a region of the Central Adriatic Sea. The approved European limit of 0.5 mg/kg wet weight for total mercury in Norway lobster was surpassed in six out of ten pools that were examined. Additionally, the higher levels of Hg concentrations in cooked samples of Norway lobster point to the need for a reconsideration of the Hg control methods now in use, taking consumer consumption habits into account. All European hake samples had Hg levels below the allowable limits and for this species, boiling had no effect on the amounts in the fish tissues.

Despite being a significant commercial and recreational fishing species in California, there is a lack of information on the bioaccumulation of the California spiny lobster (*Panulirus interruptus*) (Figure 8.6). This investigation looked at the levels of pollutants in the tissues of California's San Diego Bay lobsters. Comparing the measured lobster's tail muscle pollutant tissue concentrations to the State of California's pollution warning standards was done. After that, risk assessment was done utilizing concentrations and catch data from the California Department of Fish and Wildlife. Results of the study showed that organic contaminants did not bioaccumulate much in tail tissue, probably because there were few lipids present. Mercury was detected in amounts above-recommended levels, primarily in the methyl form. Data on the recreational catch in San Diego Bay revealed a higher non-cancer risk for fishermen whose reported yearly capture was at or above the 90th percentile.

8.6 AQUACULTURE

Aquaculture of lobster in marine cages has grown dramatically in Vietnam over the past few years. Along with providing employment opportunities for villagers in Central Vietnam's coastal regions, this sector has contributed to certain issues with the protection of the marine environment, as rubbish fish feed predominates. Using the change of productivity technique, which takes into account nitrogen release from lobster cages as a measure of seawater quality, studies examined the effects of cage marine lobster aquaculture on the environment in Vietnam. Additionally, the Mass Balance Model is used to compute the total nitrogen loading volume in each crop. There is a negative correlation between the volume of nitrogen released by a dwelling unit and the productivity of lobsters (kg/m^2). Additionally, if the number of cages in a certain location rises, lobster productivity eventually declines. On the other hand, the treatment cost calculation for this waste is used in the estimation of the environmental impact. This number, however, is less than when compared to the change of productivity technique without taking the carrying capacity of the environment into account. The Cost-Benefit analysis provides a strong recommendation to municipal authorities, marine environmental management Departments and policymakers having a better regard for the environmental and economic balance. Vietnam's marine lobster farming sector is profitable. Although there is still a marine lobster food processing sector, Vietnam's wide sea and the lack of environmental conservation awareness among lobster growers have been contributing to rising levels of marine water pollution. Therefore, it is crucial for the local authorities and any other organizations to promote the development of high-quality feed from the food industry, to have technical training programs and raise the community's awareness of environmental protection, as well as to plan strategic schemes for the entire region especially in the farming areas, in order to have sustainability in cage marine lobster aquaculture development in Vietnam.

The survival behavior of *Nephrops norvegicus* kept in experimental aquariums were found to be affected by small amounts of drilling cuttings. Exposure to cuttings had no effect on the antennule's fluttering rates or the amount of time it took to recognize and capture newly supplied food in the tanks. When the water flow via the tanks was disrupted for 12 hours, animals exposed to the greatest concentration of cuts died in 58 percent of cases, but those exposed to lower amounts survived. The survivors displayed disoriented behavior and clumsy motions once the water flow was resumed, which lasted for nearly 36 hours. Animals will be more vulnerable to predators in this situation. For native species exposed to cuttings discharge in close proximity to offshore drilling platforms, this unnatural behavior could have catastrophic repercussions.

The number of discernible impacts at a distance of 0 m and 50 m was assessed using metal compositions of sediment, lobster and feed. Sediments that were gathered beneath the cages and classified as hypoxic or anoxic had high concentrations of Cu, Zn, organic carbon and particles, but low concentrations of Mn and Fe (Farmaki et al., 2014). The influence of waste chemicals significantly decreased at a distance of 50 m. Utilizing lobster as a bioindicator species, researchers were able to identify excessive Cu accumulations at active aquaculture sites. The sediment conditions linked with the environmental monitoring program ratings were evaluated using chemical compositions and metal ratios normalized with organic carbon. Chemical data from all sites were examined using principal component analysis (PCA) to distinguish between normal, hypoxic and anoxic sediment conditions. With >90% explainable variance in the first two components, selected variables were sufficient for the PCA technique. With a few cases of EMP mis-identification, the categories based on PCA and cluster analysis were comparable to EMP classifications. The components of sediment chemistry served as reliable indicators for assessing aquaculture operating locations and marine environmental conditions. To forecast the effects of aquaculture methods and the suitability of aquaculture operations, the established methodologies, using chemical variables in combination with EMP and the statistical approach, could be helpful.

In Long Island Sound (LIS), there was a significant death of American lobsters (*Homarus americanus*) due to a number of factors. Other potential factors included immune system stress, water acidification, pesticides and microbial population shifts that may have caused shell disease. An estimated 60% of alkylphenols penetrate the marine environment where they damage the endocrine and hormone systems of vertebrates. Alkylphenols have been reported to have a variety of impacts on lobsters, including effects on larval survival, molting and postlarval molting, shell hardening and interference with metamorphosis that results in larval-juvenile intermediates. Through the use of juvenile hormone (JH) and ecdysone receptors, evidence for a fundamental molecular mechanism of action for alkylphenols was discovered.

Figure 8.7. Mediterranean mussel.

Source:https://upload.wikimedia.org/wikipedia/commons/thumb/a/a9/Miesmuscheln-2.jpg/320px-Miesmuscheln-2.jpg

Figure 8.8. Norway lobster.

Source:https://www.marlin.ac.uk/assets/images/marlin/species/web/o_nepnor.jpg

Figure 8.9. Red mullet.

Source:https://upload.wikimedia.org/wikipedia/commons/thumb/6/6f/Mullus_surmuletus.jpg/1024px-Mullus_surmuletus.jpg

Figure 8.10. Common cuttlefish.

Source:https://upload.wikimedia.org/wikipedia/commons/thumb/6/65/Sepia_com%C3%BAn_%28Sepia_officinalis%29%2C_Parque_natural_de_la_Arr%C3%A1bida%2C_Portugal%2C_2020-07-21%2C_DD_62.jpg/800px-Sepia_com%C3%BAn_%28Sepia_officinalis%29%2C_Parque_natural_de_la_Arr%C3%A1bida%2C_Portugal%2C_2020-07-21%2C_DD_62.jpg

Figure 8.11. European flying squid.

Source:https://upload.wikimedia.org/wikipedia/commons/thumb/3/3b/Todarodes_sagittatus.jpg/800px-Todarodes_sagittatus.jpg

Figure 8.12. European anchovy.

Source:https://upload.wikimedia.org/wikipedia/commons/thumb/1/17/Acciughe_2.jpg/800px-Acciughe_2.jpg

Figure 8.13. European pilchard.

Source:https://adriaticnature.com/wp-content/uploads/2015/12/Sardina-pilchardus-3-724x293.jpg

Concentrations of polychlorinated biphenyls (PCBs) and organochlorine pesticides (OCs) were discovered in the tissue of marine animals from two sites along the Abruzzo coast of the Adriatic Sea, including Mediterranean mussel (Figure 8.7), Norway lobster (Figure 8.8), red mullet (Figure 8.9), common cuttlefish (Figure 8.10), European flying squid (Figure 8.11), European anchovy (Figure 8.12), European pilchard (Figure 8.13) and Atlantic mackerel (Figure 8.14). The habitat, trophic level, feeding habits and utilization of the species in the Italian cuisine all played a role in the selection process. In order to evaluate water pollution, filter-feeding and sedentary mussels were utilized, whereas Norway lobsters were used to test sediment pollution. In the samples analyzed, PCB concentration was higher than OC levels. The pilchard had the highest levels of PCBs and OCs, whereas cephalopods had the lowest levels of both pollutants. Due to their position in the trophic chain, findings indicated that species including anchovy, pilchard and mackerel were the most polluted. The high DDE/DDT ratio indicated that DDE was more prevalent than DDT, indicating that the biotransformation rate of contaminants was very effective in fish and crustaceans. The results have also been explained in terms of the geographical distribution and biological cycle of the species.

Figure 8.14. Atlantic mackerel.

Source:https://upload.wikimedia.org/wikipedia/commons/4/48/A_mackerel.jpg

The output of suspended particles, dissolved nutrients, organic matter and bacteria from lobster cage aquaculture is the most obvious environmental impact and is assumed to be the cause of the decline in Van Phong Bay's coastal water quality. In order to address this issue, the efficacy of a co-culture system in Van Phong Bay employing *Perna viridis* (green mussel) and *Panulirus ornatus* (ornate spiny lobster) was examined. Biochemical oxygen demand (BOD), dissolved oxygen (DO), total suspended solids (TSS), nutrients and microorganisms were all measured in water samples. Bacteria and organic matter (nitrogen and phosphorus) in sediment samples were analyzed. For the safety of human health, heavy metals, pathogenic microorganisms and phycotoxins in lobster and mussel tissues were also examined. Between lobster cages with and without mussel co-culture, environmental variables in the water column and sediment did not differ substantially. TSS was, however, usually lower at sites where mussel co-culture was present. According to DO, BOD and TSS measurements, cages where lobsters were fed mussels had higher water quality than cages where they were only given garbage fish. The levels of heavy metals in lobsters fed either fresh fish or mussels were not significantly different. The findings show that mussel co-culture enhances water quality and gives lobsters a source of food. There was a limited risk to humans from heavy metals or phycotoxins while feeding mussels to lobsters.

Due to their propensity to negatively impact non-target crustaceans, anti-sea lice insecticides used in the salmonid aquaculture business are a rising source of environmental concern. However, there is little data on their effects on European lobster (*Homarus gammarus*) larvae in the Norwegian maritime environment. Azamethiphos and deltamethrin are two pesticides used on salmon farms in Norway. Potential impact zones were plotted around each farm and projected the dispersion of both substances into the marine environment around a few selected Norwegian farms using a hydrodynamic model. With LC50 and EC50 values below the advised treatment concentrations, obtained data demonstrate that azamethiphos and deltamethrin are acutely toxic to both larval stages. Additionally, it was demonstrated that while deltamethrin effect zones around farms often encompass substantially greater regions, azamethiphos impact zones were rather modest. According to these results, azamethiphos may have a less negative effect on European lobster in the Norwegian marine environment than deltamethrin.

8.7 SPECIES

Additional species of lobster, sardine, white mullet, menhaden and predatory species, such as Florida pompano, Gulf Kingfish and spot, were included in the amnesic shellfish poisoning (ASP) study. Molluscan shellfish, crabs and finfish species have all been linked to amnesic shellfish poisoning caused by domoic acid. It is most frequently linked to the eating of bivalve molluscan shellfish from North America's northeastern and north westerly shores, such as mussels, scallops and razor clams. The viscera of lobster, Dungeness crabs (*Cancer magister*), Tanner crabs (*Chionoecetes bairdi*) and Red Rock crabs (*Cancer productus*) in these areas have also been found to have dominic acid. The viscera of mackerel (*Scomber scombrus*), lobster (*Homarus* spp.), Dungeness crab (*Metacarcinus magister*), Tanner crab

(*Chionoecetes bairdi*) and Red Rock crab (*Cancer productus*) have also been found to have PSP poisons. While mackerel's viscera are typically not eaten, lobster and crab viscera could be dangerous to your health if they are taken from contaminated waters. Due to exceptionally high levels of PSP toxins, the FDA cautioned against eating American lobster tomalley from New England waters in 2008. Monitoring and necessary action should be taken in response to knowledge of emerging geographic regions and new fish species. Examples of fish species that have lately been linked to the risk of natural toxins include lionfish, which have been found to have high amounts of CFP that can make people ill and lobster, notably the tomalley, which contains PSP and ASP.

In the event that a processor decides to install controls somewhere, those controls must offer a comparable level of safety assurance and should be backed by reliable scientific data. The poison from the ingested portion of fish or molluscan shellfish may occasionally be able to be removed through processing. These exclusions depend on the toxin, procedure and kind of fish or molluscan shellfish. Examples include removing the adductor muscle from scallops or eviscerating the fish, such as lobsters, crabs and anchovies. During aquatic holding of live lobsters and crabs in tanks or the shipping of live fish, it would be reasonably possible that dangerous amounts of drug residues from aquaculture would be introduced. In order to store live lobster in a lobster pond, a primary processor must utilize oxytetracycline as a medicated feed in accordance with labeling instructions and must record the withdrawal period of 30 days prior to selling. The CCP should be set by the processor to hold. These goods may be prepared by the processor, acquired previously prepared, or put together by the processor from ready-to-eat parts. They might look to the customer or end-user to be products that are ready to eat, in which case they could be consumed without further cooking. Examples include hot-smoked fish, seafood salads, cooked crabmeat, lobster meat and crayfish meat. Surimi-based analog items are another example.

In some instances, a secondary processor will accept chilled, cooked, ready-to-eat meals such as lobster meat, pasteurized crabmeat, smoked fish and surimi-based analog goods and hold them for sale without additional processing. In other instances, a secondary processor will receive these goods and use them as ingredients in a ready-to-eat product without cooking it or pasteurizing it. To decrease the risk of the growth of harmful bacteria in these circumstances, the receipt and storage processes by the secondary processor should be marked as CCPs. The receiving and storage steps prior to the cooking or pasteurization step may not need to be designated as

CCPs, however, if the secondary processor receives the ready-to-eat foods to be used in a product that will be cooked or pasteurized by that processor, provided that the formation of *S. aureus* or *B. cereus* toxin is not a significant risk. Keep in mind that heat is unlikely to render these poisons inactive.

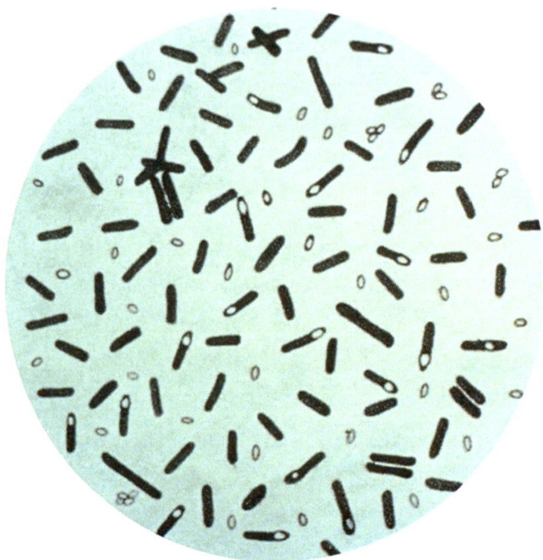

Figure 8.15. *C. botulinum.*

Source:https://upload.wikimedia.org/wikipedia/commons/5/58/Clostridium_botulinum_01.png

8.8 CONTAMINATION

In order to reduce the risk of recontamination with pathogens like *C. botulinum* (Figure 8.15) type E and non-proteolytic types B and F, the product must be filled into the final container while it is still hot in a continuous, closed filling system. Products like soups, chowders and sauces that are poured directly from the cooking kettle, where the risk of recontamination is reduced, are covered by this control technique. Items that are handled between cooking and filling, such as crabmeat, lobster meat, crayfish meat and other products, might not be covered by it. The only defense against *C. botulinum* type E and non-proteolytic types B and F toxin production during complete product storage and distribution is refrigeration. This is true for refrigerated, reduced oxygen packaged raw, unpreserved fish such as refrigerated, vacuum-packed fish filets and refrigerated, reduced oxygen packaged, unpasteurized, cooked fishery products (such as refrigerated, vacuum-packed crab meat, lobster meat, etc. These *C. botulinum* varieties can survive in environments as cold as 38°F. Post product leaves your control and enters the distribution system, it is typically impossible to maintain temperatures below 38°F, as was previously mentioned. To address these issues in the distribution system, it may be suitable to apply a TTI on the smallest unit of packaging i.e., the unit of packaging that will not be distributed any further, typically a consumer or end-user package(Nigam & Nigam, 2010).

Cooking is a heat-treatment process that is often carried out before the product is put in the container for the final product. It is used to describe fishery products that are sold frozen or chilled. Fishery products are typically described as prepared, ready to consume after cooking. Crabmeat, lobster, crayfish, cooked shrimp, surimi-based analog products, seafood salads, seafood soups and hot-smoked fish are a few examples of cooked, ready-to-eat fishery products. To avoid melanosis, or black patch, on shrimp and spiny lobster shells, sulfiting agents are frequently used as preservatives. Additionally, they can be used to prevent conch meat from browning, keep the red color of the octopus' skin when the animal is cooked and as a breeding component. When the concentration meets or surpasses 10 ppm, the FDA requires processors to disclose the presence of sulfites. If a sulfiting agent is present in the food, its use and/or concentration will influence whether or not it is listed as an ingredient on the label.

The ocean provides vital services like nutrition, climate regulation, coast protection and pollution absorption. This blue economy's size formerly created the appearance that services were endless, but decades of overuse and widespread contamination have resulted in habitat deterioration and lost ecological function. Due to this, marine ecosystems are less able to continue providing these essential services for human survival. The burden on seas to provide essential ecosystem services like food, climate management and storm protection increases as the global human population rises to the estimated 9 billion by 2050. Currently, one billion people depend solely on seas for their nutrition. It is expected that this will last to rise, especially in developing nations like the Republic of South Africa where inequality and food insecurity are increasingly pervasive. South Africa has made political and economic progress, but inequality, a stagnating economy and high rates of poverty still plague the nation. South Africa has committed to expanding its economy, promoting environmental sustainability, eradicating poverty and reducing inequality in the fishing industry by 2030 in accordance with the UN-backed Sustainable Development Goals (SDGs). Decision-makers need cutting-edge research that indicates potential future changes affecting the achievement of these SDGs in order to reach them. Ocean research with an SDG focus is urgently needed because 3 billion people currently rely on marine ecosystem services for their livelihoods.

One study strategy for sustainable ocean management that has grown in prominence in recent years is the valuation of ecosystem services. The advantages that humans receive from ecosystems are referred to as ecosystem services. Four interconnected categories like provisioning, sustaining, regulating and cultural services can be made up of these services. Decision-makers can better integrate environmental, social and economic issues into policy and management by valuing ecosystem services. An economic approach that ignores natural capital is detrimental to the earth and, consequently, our own long-term destiny given that there are around 2 billion people living in poverty. Decision-makers can adopt a uniform gauge for both the tangible and intangible advantages emerging from various management scenarios by including natural capital into economic appraisals. Users can model spatially explicit outcomes of future scenarios based on ecological production and anthropogenic pressures using some valuation methodologies, such as the Integrated Valuation of Ecosystem Services and Trade-offs (InVEST) methodology. Decision-makers can connect the direct advantages, synergies and trade-offs related to various management initiatives by valuing the future services offered by nature. Changes in inequality, poverty, environmental sustainability and economic growth may be among these advantages, synergies and trade-offs. Decision-makers may make educated, solid judgments to guarantee the SDGs are achieved.

CHAPTER 9

Challenges in Lobster Culture

CONTENTS

9.1 INTRODUCTION

The successful production of spiny lobsters in captivity has been reportedly hindered by slow growth, growth discrepancy and growth depensation. The relationship between individual metabolic rate variance and behavior and growth has not previously been studied in any spiny lobster species, despite earlier research with a variety of aquatic ectotherms suggesting that it may be a factor. Social connection is critical for fostering lobster growth, as seen by the much greater growth performance, survival rates and feed intake in communal rearing.

Standard metabolic rate, typical metabolic rate and growth were positively correlated in lobsters raised alone, proving a causal association between metabolic phenotype and lobster growth in the absence of social contact. In communal raising, the impact of social contact overshadowed the direct correlation between metabolic rate and lobster growth. The findings show that the ability of spiny lobsters to grow is associated with individual differences in metabolic status, with social behavior having a significant impact on individual growth. The aerobic energetic cost of living, or metabolic rate (MR), is what is needed for an organism to perform, process and function in order to sustain life. In aquatic ectotherms, this baseline energy expenditure is known as standard metabolic rate (SMR). Routine metabolic rate (RMR) is the average metabolic rate while an animal is not moving much, whereas active metabolic rate (AMR) is the maximum pace at which aerobic energy metabolism can occur. The total quantity of aerobic energy accessible to the animal can be calculated by deducting the minimum from the maximal MR. A crucial physiological characteristic that affects an organism's performance is metabolic rate. Pelagic fish and other active organisms typically have greater MRs.

For some fish and crustaceans, MR may be heritable and reproducible throughout time. The rate of energy metabolism, or metabolic phenotype, can also vary greatly among individuals within a species. The mechanisms that can affect an organism's fitness, such as growth, reproduction, survival, dominance and life duration, have been connected to metabolic characteristics. Individual food intake and growth may be impacted by individual metabolic rate, which may also affect locomotor abilities, dominance and aggression. However, depending on the surrounding environment, the link between metabolic phenotype and fitness can vary. Individuals with a high SMR may benefit from living in a setting with plenty of food, such as in captivity. On the other hand, under unfavorable circumstances, such as a food shortage, a low SMR can be advantageous. Spiny lobsters have a reputation for being aggressive and possessing complex social behaviors that they use in the wild to establish social hierarchies, choose shelters, avoid predators and even develop behavioral immunity (Iv, Jr, Iii, Moss, & Shields, 2015). Slow growth, significant growth disparity and depensation have been identified as the main obstacles to the successful production of spiny lobsters in captivity. Agonistic behavior and hierarchical social structures, in which the dominant individuals control a disproportionate amount of the food resources, are frequently connected with the growth depensation in spiny lobster rearing. Individuals develop at varying rates as a result of dietary variances, which can over time increase the variance of size distribution.

Prior research has not been done on the association between the metabolic phenotype, dominance behavior and growth of spiny lobsters during raising. New developments in hatchery production have led to the emergence of the aquaculture species *Sagmariasus verreauxi* (Moss et al., 2004). To ascertain the potential impact of social behavior on the link between metabolic phenotype and growth performance, juvenile lobsters from hatcheries were reared either singly or in groups. It was expected that growth disparity and dominance hierarchies are both influenced by the metabolic phenotype, with individuals with higher metabolic rates being more dominant and developing more rapidly during communal upbringing. Finding the root reasons of growth discrepancy and depensation in rearing, which are major obstacles to production consistency, may need a better knowledge of the link between metabolic phenotype and growth.

9.2 GROWTH

Growth, molting and eating behavior in *S. verreauxi* juveniles were significantly impacted by rearing methods. Lobsters raised communally grew noticeably more quickly, with higher size differential and growth depensation than those raised individually. Similar to this, lobsters of *Panulirus ornatus* grew 40% more quickly when raised collectively than when raised separately. Spiny lobsters have been found to exhibit aggressive behavior and the capacity to establish highly organized dominance hierarchies. By encouraging direct food rivalry, appetite suppression, decreased food-conversion efficiency and an increase in motor activity, the presence of dominant species can either directly or indirectly inhibit the growth of others. Agonistic behavior in spiny lobsters resulted in size disparity and growth depensation because the dominant individual had a disproportionate amount of control over food supplies relative to the others. Communal growth resulted in a higher CVWG, showing that communal lobster culture appeared to accelerate growth depensation in the early phases of juvenile development. Adults of *Jasus edwardsii* did not exhibit any growth depensation when raised over an 8-month period suggests that the growth depensation of *J edwardsii* may diminish with aging and proper nutrition. Studies did not find any evidence of cannibalism or mortality in communal rearing, proving that these characteristics were not responsible for the communal system's superior development. Over the second molt, the intermolt period was noticeably shorter in lobsters raised communally than it was in those raised individually. This outcome is in line with findings from *P. homarus* and *P. ornatus*, which showed that communally reared lobsters molted more frequently than those raised separately. Lobsters can detect the scent of conspecifics, including chemicals released during molting, agonistic behavior and reproductive behavior. A chemical signal from molting conspecifics may be crucial for stimulating molting, which may help to explain why lobsters raised separately underwent longer molts. Lobsters raised communally showed higher AFI, which may have helped them grow more quickly. The competitive interactions among the lobsters in the communal growth may have caused the triggers or cues to stimulate the lobsters' feeding. Highly sociable species, spiny lobsters are known to congregate in large groups both in the wild and during the rearing process. This may be crucial for managing stress and fostering optimal feeding and growth when conspecifics are present. Controlled trials have demonstrated unequivocally that social interaction affects crab growth.

9.3 INTERACTIONS

Social interaction affected the direct correlation between growth and metabolic rate in communal raising. Due to the energetic cost of dominance, high SMR lobsters may not profit from the higher intake concept as seen in individual raising lobsters. As a result, it might benefit juveniles with lower metabolic rates by enabling them to grow as quickly as the dominant in communal growth, as indicated in the compensation hypothesis, where more spare resources can be focused on growth. The influence of sex, dominance and competitive aptitude are just a few factors that could nullify the growth advantage of the high metabolic rate of animals in community rearing. The findings indicate that, in neither type of rearing environment, growth and sex do not interact. This may be as a result of the analyzed lobsters' immaturity. Animals with high SMR can be very aggressive, according to earlier research on aquatic invertebrates. Additionally, being overly aggressive can lead to serious expenditures such as energy used during agonistic competitions or missed feeding opportunities. Therefore, the costs of aggression may indirectly lead to a non-significant link between metabolic rate and growth.

According to the findings, there is no correlation between individual-rearing metabolic rates (SMR, RMR, AMR and AS) and AFI in spiny lobsters. Studies found no correlation between individual AFI and metabolic rate, which suggests that lobsters with faster metabolic rates may not need to consume more to cover their higher maintenance costs. This might be a result of having better digestion and absorption efficiency than lobsters with lower metabolic rates. Individual conversion efficiency is strongly associated with physiological and anatomical characteristics, as well as feed consumption. Studies on rainbow trout showed that the size of gastrointestinal tracts, maximum eating capacity and growth efficiency are more closely related to fish growth rates than their metabolic phenotype. The likelihood of inaccurate AFI measurement, which is known to have inherent imprecisions in studies with crustaceans due to their messy eating habits, may potentially be responsible for these results. Spiny lobsters are classified as messy eaters because they use their mouthparts to shred, pull and grind food into smaller pieces before eating it.

Spiny lobsters' growth in captivity is significantly influenced by their social behavior. Competition for scarce resources, like food, in a group of social animals has a significant role in how individuals develop. The mix of an animal's morphological and physiological traits, such as body size, sex and energy reserves, as well as previous experience, frequently determines how successfully it can compete for resources. These studies looked at how factors such as lobster size, metabolic phenotype, sex, participation in feeding competitions and rearing history affected the social dominance status of juvenile spiny lobsters (*Sagmariasus verreauxi*). Competitive ability and dominance demonstrated by the lobsters were measured by recording orientation, avoidance, fighting and eating actions. The social dominance status of the spiny lobster was significantly predicted by size. Furthermore, sex, size and metabolic phenotype were all substantially correlated with social dominance position. Larger lobsters were projected to be dominant over smaller lobsters, while female lobsters, regardless of size or metabolic phenotypic status, showed the potential to become more dominant than male lobsters. In distinction to what has been observed with fish and other species of crustaceans, low metabolic rate lobsters showed a greater capacity to outcompete high metabolic rate lobsters. These results show that a combination of metabolic phenotype, sex and individual lobster body size can be employed as reliable predictors of social dominance status in spiny lobsters. Therefore, research implies that particular morphological and physiological characteristics have a significant influence in determining the social behavior status of spiny lobsters and, consequently, growth performance.

Conflict and subsequent violence frequently play crucial roles in establishing the distribution and structure of an individual animal in a group or population, such as the establishment of social dominance hierarchies. Social dominance hierarchies can be used to determine species status and ability to obtain food and shelter. In aquatic species, differences in size, sex, molting stage, feed availability, recent agonistic experience and energy metabolism are all substantially correlated with the dominance position and intraspecific competitive abilities. It is crucial to identify the traits of individual animals that affect how agonistic interactions turn out in order to forecast an animal's social standing and rate of growth. Larger males are typically more aggressive and tend to dominate over smaller conspecifics, frequently monopolizing resources like food and shelter. This is true for most crustaceans. Larger lobsters can frequently maintain their size status in the spiny lobster *Jasus edwardsii* when eating possibilities are more constrained, but as feed supply grows, less of the originally largest size-ranked lobsters can do so. In the giant freshwater prawn *Macrobrachium rosenbergii*, metabolic phenotype has also been documented to influence dominance status. Individuals with higher energy metabolism were found to be more aggressive and dominating than others.

In the wild and in captivity, spiny lobsters frequently live and seek shelter with other conspecifics, exhibiting sophisticated social behavior. Spiny lobsters have aggressive and complicated social behaviors that they employ to create social hierarchies. Dominant individuals are able to control a disproportionate amount of food resources in comparison to others. Previous studies showed that *Panulirus ornatus* and *Panulirus homarus* juveniles showed a bigger development rate and growth disparity than individually reared juveniles when social contact was associated with communal raising. *S. verreauxi* has also shown comparable patterns. Greater feed intake during communal rearing than during individual rearing is thought to be associated with improved growth performance. This was probably brought on by social interactions, such as feeding competitions, with conspecifics.

As a result, communal parenting will experience larger growth disparity and, consequently, growth depensation. However, there was no correlation between the metabolic phenotype and growth when the social engagement was taken into account. To understand the apparent mismatch between energy metabolism and growth in communal culture, more knowledge of the relationships between metabolic phenotype and social behavior is needed.

9.4 MORPHOLOGY

Size and sex are the morphological features and metabolic phenotype is the physiological trait. It may be possible to determine the root of the discrepancy between energy metabolism and growth in communal raising and growth performance of individual lobsters in captivity by comprehending the consequences of these variables on social dominance status and feeding competitive behavior. Threats, aggression and submission are all examples of agonistic behavior in social interactions between species who share resources, primarily food and shelter and which may have both direct and indirect consequences on things like damage, mortality and growth. According to earlier research, spiny lobsters exhibit antagonistic interactions, particularly while maintaining social hierarchies like shelter and food hierarchies. The approach of one animal, typically more dominant, towards the other typically starts an aggressive interaction between spiny lobsters. Upon the arrival of food, dominant and neutral lobsters displayed aggressive behavior by launching attack behavior. Additionally, dominating lobsters spent more time fighting, assaulting and feeding. Similar to behaviors documented in other Palinurid species like *Panulirus cygnus*, the dominant lobsters demonstrated agonistic behavior.

Female lobsters have the ability to become more dominant than male lobsters, regardless of their size and metabolic phenotypic status, according to a substantial relationship between social dominance status, metabolism status and sex. Larger size lobsters are more likely to become dominant than smaller size lobsters, with the dominance index rising as the size difference between pairs increases. All the models showed a significant positive relationship between lobster size and dominance status, indicating that size is an important predictor of spiny lobster social behavior status. Previous research on crustaceans has demonstrated that body size is one of the key factors in predicting the outcome of competition. Similar research on young *Cherax cuspidatus*, freshwater crayfish showed that dominant status was established depending on individual animal body size, with larger individuals more likely to win a pairing contest than smaller individuals. Individual metabolic rate variation has been extensively researched in relation to aggressive behavior and dominance hierarchy in fish and some crustaceans.

With 11 molt stages, the southern rock lobster, *Jasus edwardsii*, may have one of the longest larval phases of any marine animal (Breen, 2009). The emerged juvenile can show significant body size variation after metamorphosis, which may be related to their experiences as larvae. It is well recognized that the key determinants of individual variance in growth performance of juvenile aquatic invertebrates are the larval experience, size and condition of post-metamorphosis individuals. The impact of emergent juvenile size on individual spiny lobsters' subsequent growth performance, however, has not been studied. In a lab experiment, the impact of body size, metabolic phenotype and social behavior on *J. edwardsii* emerging juveniles' growth increments was investigated. The lobster's performance in terms of growth was improved by this proposed social interaction factor. The emergent juveniles' body sizes showed no discernible effects on growth rates in either raising strategy. While communally reared lobsters showed no correlation, lobster growth in the individual rearing treatment was positively correlated with routine metabolic rate, active metabolic rate and aerobic scope. This suggests that metabolic phenotype is a key factor in determining growth in the absence of social interaction. The findings showed that the ability of spiny lobsters to grow is associated with each individual's metabolic phenotype, with social behavior having a significant impact on each lobster's growth. A widespread species of temperate spiny lobster found in the coastal waters of southern Australia and New Zealand is the southern rock lobster, *Jasus edwardsii*. This species has been recognized as a potential aquaculture species and is one of the most valuable fishing resources in Australia and New Zealand. *J. edwardsii* demonstrated size discrepancy among individuals from the same cohort under aquaculture circumstances, similar to the majority of spiny lobsters, as a result of the variation in individual growth performance. The elements that may affect spiny lobster growth performance in captivity have been the subject of research up to this point.

J. edwardsii exhibits a lengthy larval life phase, similar to other spiny lobsters. Its free-swimming planktonic larvae, called phyllosoma, spend 8 to 24 months in ocean water before undergoing metamorphosis into the puerulus stage, which migrates back to shore to find an appropriate benthic settlement substrate (Ritar, 2001). The puerulus is a non-feeding stage that meets its energy needs by drawing on reserves built up during the phyllosoma phase before it. The puerulus settles and completes its transition into the emergent juvenile stage. Emergent juveniles of *J. edwardsii* can have a wide range of body sizes after metamorphosis, with post-puerulus ranging from 10-15 mm in carapace length (CL) (Crear et al., 2000). The disparity in emergent juvenile size may be due to individual differences in the environmental and nutritional conditions they experienced during the larval development period, which could be related to the inter-individual variation in the emergence juvenile condition. It has been understood that changes in embryonic or larval experiences can account for within-species diversity in the juvenile performance of aquatic invertebrates. Latent effects play a role in the individual variation of a wide range of performance parameters in many aquatic species, including growth rates, survival, competitive ability and environmental tolerances. One could anticipate that latent effects would have a significant impact on the species given the lengthy period and intricacy of spiny lobster larval development.

9.5 METABOLIC RATE

To comprehend the fitness of the organism, knowledge of the effects of metabolic phenotype on individual animal growth performance is crucial. The metabolic rate is a crucial physiological characteristic since an organism's ability to operate and function aerobically is essential for maintaining life. Additionally, it is thought that the role of metabolic phenotype in describing the patterns of adaptation and the variety of life histories found in different animals within a species. Individual differences in metabolic phenotype have been linked to differences in growth, reproduction, survival, lifespan and behavior in fish and crustaceans. It is also widely known that the metabolic phenotype of populations can vary significantly amongst individuals and this might correspond to variations in food intake, growth and behavior. There is evidence that aggressive and dominating behaviors play a significant role in spiny lobster aquaculture. Aquatic species' individual behavioral variance may be influenced by their metabolic phenotype. Higher metabolic rate people can be more aggressive and develop faster, giving them a size advantage that secures a higher place in a social hierarchy. Similarly, research on crustaceans has demonstrated that social interaction and metabolic phenotype might influence growth and survival in rearing conditions. Early research has shown that social groupings of animal species are frequently separated based on morphological phenotypic features, such as size, age and sex. Additionally, social animals frequently exhibit wider behavioral variations within a cohort, with some individuals constantly being more exploratory, aggressive and energetic.

Previous research has shown that variations in metabolic physiology can affect how differently spiny lobsters grow, with higher metabolic phenotypes seeing increases in carapace length and molt frequency when social interaction is absent and food is abundant. However, it is unclear how metabolic phenotype and growth are related. Studies on teleosts indicate that food availability and intake have an impact on the link between growth and energy metabolism; however, studies on spiny lobsters have not looked at the association between metabolic phenotype and feed consumption. First (J1) and third (J3) instar juvenile stages were used to investigate the apparent feed intake (AFI) and food preferences of emergent juveniles fed with three different feeds: mussel gonad (MG), mussel mantle (MM) and a commercial diet (MFD). The findings showed that neither the metabolic phenotype of a human nor the growth performance of a lobster was connected to feed consumption. The preferred food of young lobsters was mussel gonad and food preference was not associated with individual growth performance. These findings suggest that feed intake and desire are not basic determinants relating to metabolic phenotype and growth performance. Instead, a complex network of intrinsic and extrinsic factors may be involved, which calls for further investigation. Due to over-exploitation of the wild stock, spiny lobsters are profitable, in high demand and have a limited supply. Spiny lobsters have been cultivated in aquaculture using either hatchery-produced or wild-caught juvenile seed stock in an effort to meet both the demand for their products worldwide and the fishing strain on wild stocks. The southern rock lobster, *Jasus edwardsii*, is one of the most profitable temperate species of spiny lobster. It is also a viable prospect for aquaculture, notably through the on-growing of wild juvenile seed. Large inter-individual growth rate variation, which causes growth disparity and ultimately lower lobster biomass, remains one of the main obstacles to the effective breeding of spiny lobsters in captivity. Due to the variation in individual development performance, *Jasus edwardsii* in particular has shown significant size discrepancy across individuals within a population.

Spiny lobster populations exhibit size variation among individuals beginning in the early life stages and as the lobster matures, this variety becomes more pronounced due to variations in individual growth rates. Early studies on *J. edwardsii* revealed that differences in nutritional status of individuals upon completion of the larval stages may be related to inter-individual size variance in post-puerulus; for example, larger pueruli had higher energy stores. The emergent juvenile, also known as first instar juvenile (J1), of J. edwardsii revealed significant variances in size, ranging from 9.3 to 12.8 mm carapace length (CL). The size variance of the individually reared fifth instar juvenile (J5) increased as the emergent juveniles matured, with sizes ranging from 18.8 to 27.4 mm CL (J. C. Groeneveld, Greengrass, Zyl, & Branch, 2010). The fact that the variation in subsequent individual growth performance was unrelated to emergent juvenile size suggests that additional factors were involved in the diversity in individual growth performance. It was also demonstrated that, in the absence of social interactions, individual variation in *J. edwardsii* growth was connected with metabolic phenotype. However, the underlying mechanism connecting metabolic phenotype with growth performance has not yet been clarified. Early juveniles of *S. verreauxi* and emergent juveniles of *J. edwardsii* were used in the experiments and it was discovered that individual variation in metabolic phenotypes could affect growth differences in both species, with higher metabolic phenotypes having greater growth performance than lower metabolic phenotypes. These findings are the first to directly state a connection between individuals' variations in metabolic phenotypes and crustacean development performance. The outcome also compares strongly with earlier teleost investigations, which strengthens the significance of metabolic physiology in determining the growth of specific aquatic animals. Additionally, this data implies that metabolic physiology is a key element influencing growth in juvenile spiny lobster culture, whether it is from hatcheries or wild stocks.

Previous research on teleosts revealed that the availability and consumption of food affect the link between energy metabolism and growth. Research suggests that metabolic characteristics were not associated with individual lobster meal intake in spiny lobster. There was no relationship between lobster feed intake and growth performance. *Jasus edwardsii* emerging juveniles showed a preference for mussel, namely mussel gonad, over a prepared meal, which is consistent with other investigations. However, there was no correlation between food preferences and individual growth performance. Overall, these results indicated that neither feed intake nor food liking are critical determinants of growth success in lobsters. Therefore, superior feeding efficiency, nutrient intake, or other factors such as individual digestive and anabolic capacity are vital as proven in fish. Since spiny lobsters are a very social species, previous research has shown that they frequently congregate with one another in shelters. The effect of social interaction on spiny lobster growth performance was examined. The results demonstrated that social behavior had a significant impact on the growth of *S. verreauxi* and *J. edwardsii* as well as on growth disparity and depensation in populations. Results also showed that the direct relationship between metabolic rate and lobster growth was overwhelmed by social interaction, indicating that social behavior governs population increase more predominantly than metabolic rate. Both of the examined species demonstrated that social engagement had a positive impact on lobster growth performance and feed intake. These outcomes are in line with earlier research on *Panulirus ornatus* and *Panulirus cygnus* and they lend further credence to the idea that communally reared lobsters exhibit better growth performance than individually reared lobster, which is a crucial factor for commercial aquaculture. Cannibalism may be a factor in the reduced survival of *Panulirus ornatus* in communal cultures, but in the current study, survival rates for *S. verreauxi* and *J. edwardsii* were high and there were no signs of cannibalism (Jones, Linton, Horton, & Bowman, 2001). The findings might be explained by variations in social behaviors of *S. verreauxi* and *J. edwardsii*. Before *Sagmariasus verreauxi* was divided into an older monotypic genus Sagmariasus, *Sagmariasus verreauxi* and *J. edwardsii* were both originally part of the genus Jasus. In contrast to *Panulirus ornatus*, which has been described as a more solitary and aggressive species, especially during early juvenile growth, both species are known to be highly sociable. This suggests that Australian temperate species may be simpler to cultivate to market size than tropical species, if growth rates can be enhanced. The increased feed intake also provided evidence that social engagement may operate as cues or triggers to encourage juvenile feeding, presumably in reaction to competitive interactions.

To identify the processes, such as visual, chemical, and/or physical association, which encourage improved growth of spiny lobsters in community culture, more investigation is needed. The dynamics of social interaction in crayfish and lobsters have been shown to be influenced by the usage of visual and olfactory signals. According to reports, lobsters can detect conspecific odors such as those released during molting, agonistic behavior and reproductive behavior. Additionally, attraction to chemosensory cues from conspecifics facilitates the sociable traits of spiny lobsters. A key component of the behavioral range before, during and after contact is the communication between individuals within a community. *Orconectes rusticus* crayfish research has shown that chemical cues from individuals are enough to stop any further interaction from occurring between individuals. Future research on spiny lobsters is necessary to identify the behavioral and sensory mechanisms that underlie the apparent acceleration of growth in connection with social interaction in communal cultures. Studies have demonstrated how complex and influenced by a variety of internal and extrinsic factors the growth of spiny lobsters in culture is. Collectively, the findings show that social behaviors are more essential than metabolic characteristics in supporting the growth of populations. When social connections were lacking and there was plenty of food, metabolic physiology had a more direct impact on lobster growth performance. Experiments with feeding competitions demonstrated that lobster body size, metabolic characteristics and sex can all affect the dominance status of spiny lobsters during eating. These findings imply that the relationship between individual spiny lobster physiological features and social interactions can affect the growth performance of spiny lobsters in culture. The analysis of the effects of initial individual diversity in body size and feed intake indicates that neither of these variables is a significant contributor to growth differences in culture.

9.6 POPULATION

As the global population grows, so does the demand for fisheries resources and the sustainability of all fisheries depends on the acceptance and implementation of proper management regulations that safeguard the breeding stock and preserve the habitat in which they thrive. Intelligent management practices should always be the first option when contemplating how to raise or maintain stock levels because they are essential to the self-sustainability of any lobster fishery. The preservation and self-sustainability of a fishery depend on the avoidance of netting adequate quantities of ovigerous or egg-bearing female lobsters. To support stocks and increase the economic benefits from fishing, a variety of technical techniques can be implemented. Any management policy must also have dynamic content that can change based on the situation. There have been numerous documented attempts to raise clawed lobsters on both sides of the Atlantic, with varied degrees of technical success and failure. Numerous issues regarding the practicality and effectiveness of raising juvenile lobsters for release into the wild remain unanswered as a result of these efforts. The production of young lobsters via aquaculture methods and the subsequent release of such lobsters into the wild have the potential to lead to irrational expectations about potential returns. Therefore, a thorough assessment of the possibility of stock augmentation should be made before starting such an effort. This should involve developing a scientific monitoring program, comprehending the life cycle of the target species and estimating costs economically. After conducting such an examination, if it is determined that the project has promise, a thorough experimental plan that takes into account all of the goals, expenses and scientific factors should be created before work begins.

Numerous widely accepted terminological schemes have been applied to the life cycles of *H. gammarus* and *H. americanus*. Despite their many similarities, discrete life stages of *H. gammarus* exhibit different behavior in the field than their American equivalent (Carlberg, Van Olst, & Ford, 1978). The three free-swimming or pelagic stages of lobster larvae's development after hatching from eggs are symbolized by molting. In a hatchery setting, this process can take 10 to 14 days depending on temperature, while it is believed to take around a month in the wild. These three free-swimming larval stages move by using their thoracic appendages, known as pleopods and expodites. Plankton frequently contains both Homarus species' larvae up to Stage IV post larvae. Stage IV post-larvae dive out to the benthos to survey the bottom and then decide whether to settle there or swim to a more favorable area for habitation. These benthic or established lobsters have been found in North America. Despite rigorous scientific surveying, no new recruits or juvenile lobsters with a carapace length (CL) less than 20 mm have been found in European seas so far. Biological principles underlie the majority of aquaculture shellfish and marine finfish hatcheries, albeit the tools and methods may differ. The methods used in this article's hatcheries are a mash-up of approaches that have been modified for raising young lobsters. Many concepts have been taken from other sources and changed to fit the operational needs of this inquiry.

Although cleanliness and hygiene should be followed in effective hatchery operations, there are so many variables that lobster production is often more of an art than a precise science. To perform menial duties like cleaning, an operator may need sound judgment, zeal, experience and dedication. These characters are difficult to teach and are typically learned through experience. Experience has shown that raising larval lobsters during the clawed lobsters' natural breeding season is most effective since survival rates are higher and production costs are lower. The additional benefit of running a hatchery seasonally is that other marine species can be produced there while the generation of lobster larvae is not going on. Each of the Stage IV lobsters produced had to be removed from other more immature stages and then placed into individual compartments before being counted. By raising the temperature of the water, the broodstock was kept in from

January to May, early hatchlings were produced, while later hatchlings were somewhat produced by cooling the water in October. Many operators still choosc pump-operated Kreisel systems because they have historically been used to raise lobster larvae for stock improvement. The average survival rate with this kind of system employing foods could be poor.

9.7 FEED

It involves enriching Artemia with two phytoplankton types and feeding the enriched Artemia to lobster larvae in conical vessels. Following this, parts of the methodology employed in Maine were adopted and adapted in Ireland to meet the needs of the European lobster *H. gammarus*. When the hatchery was first created, it was discovered that larvae had an average survival rate of about 40% during the typical hatching season from Stage I to Stage IV after 11 to 14 days in the hatchery (Militz, Leini, Duy, & Southgate, 2018). The broodstock holding, hatching containers, larval rearing hoppers, algal culture and nursery systems including air temperature control are all parts of the lobster production cycle used for stock enhancement. These stages resemble the operation of a bivalve shellfish hatchery and adequate biological quality saltwater is crucial for successful hatchery operations. Another requirement is that the used equipment be easily accessible and economical to run. With the exception of a UV system, an air blower and seawater supply pumps, expensive equipment was avoided in this project. Following the selection of the broodstock, they were weighed and placed in plastic fish boxes with their tails tucked under them to save their eggs. They were wrapped in damp bags or sponges and the box's lid was made of polystyrene. During transit, it is crucial that they are kept cool, moist and calm and that the trip lasts as little time as possible. If not, the eggs will stop being viable. When they arrived, lobsters were separated into several 20-liter perforated black barrels based on how far along the egg development process they were in. Each drum featured a concrete base that served as a weight and prevented the lobster from turning the drum on its side and escaping. These bases also had wide ventilation slots for good water exchange. These shelters provided excellent low light conditions for lobsters, which are solitary crevice-dwelling species. It was discovered that egg loss and mortalities were significantly lower in these shelters than in lobsters held communally.

According to the level of egg maturity in raceways, the drums carrying the lobsters were stacked in order. Each raceway was continuously fed with ambient, unfiltered saltwater and vigorously aerated at various locations.

The broodstock was periodically checked for egg development and kept in the ponds until it was transferred to conditioning systems or the hatchery. Monitoring egg maturity helped the hatchery to maintain a steady supply of larvae. If they were to be kept for a long time, broodstock would be fed a range of meals. Depending on what was available, they typically ate fish. These lobsters were not fed for at least two weeks before the eggs were scheduled to hatch. It seemed that keeping broodstock in the hatchery for an extended period of time was harmful to the survival and development of the larvae and it was best to buy freshly caught lobsters with eggs that were almost ready to hatch rather than feeding the broodstock.

The broodstock was relocated into an adjacent, clean raceway without being disturbed during routine raceway cleaning. Planning for postlarval production must take ovigerous lobster fecundity into account. Female lobsters kept in captivity have been documented to experience variable egg loss. The goal of the management philosophy mentioned was to produce well-fed, relaxed animals that are less inclined to fight or steal eggs from their brood. The strict application of hygienic techniques for the development of larvae is one of the key requirements for generating Stage IV lobsters and operational procedure complacency was the biggest threat to the production process. Every employee that works in the hatchery adheres to basic but crucial hygiene procedures. For instance, footwear was cleaned in disinfectant foot baths at the hatchery's entrance. The operators cleaned their hands before beginning work and put on rubber gloves thereafter. In the hatchery, the gloves were also washed in between each task. Every person entering the hatchery must wear clean attire. Additionally, it is advised that workers start in the hatchery and complete all of their tasks there before handling animals outside the hatchery or seawater. The hatchery should be kept isolated from any potential contaminants, which is the most crucial step.

Depending on how many eggs are carried, a single female lobster's hatching phase can span anywhere from three to fourteen days. Larger females produce more larvae and take longer to release all of them than smaller females. The number of larvae discharged once hatching starts often peaks between days 2 and 6 in the hatching arteries and is usually finished by days 8 to 14. Occasionally throughout this investigation, daylight larval hatching was observed. This was assumed to be caused by stress during the transfer of broodstock to a new hatching tank and temperature changes of the fresh seawater. When the larvae were examined, there were no obvious abnormalities. However, on other occasions, tiny premature larvae were produced. In the winter (January/February), when the photoperiod was reversed for two to three weeks per individual lobster, daylight hatching was also observed. When an egg hatches, internal pressure ruptures the egg's outer membrane, causing it to peel off and the inner membrane to split. The stream created by the beating pleopods carries the larvae away. The pre larva immediately performs a molt to become the first larval stage after hatching. The larvae transition to a planktonic state and float to the surface.

It's critical that the lobster hatchery is well-insulated and segregated from any potential disease reservoirs. The building should be able to sustain constant air temperatures throughout the winter and spring while the larvae are growing. The fecundity of the European lobsters was too low outside the main hatching season for these vessels to be operated efficiently, so white 90-liter polyethylene hoppers were mounted on tables inside the hatchery. Aeration tubes were inserted through the base of each hopper to provide large air bubbles. Simply pushed through a hole that has been bored slightly smaller than the airline's diameter, the 5 to 8 mm airlines are connected to an air distribution pipe that is located above the hoppers. In order to avoid larvae and algae from backing up into the airline in the event that the air blower fails, it is crucial that the primary air distribution pipe is located well above the hoppers. Airflow to the hoppers was controlled by clips fastened to the airline. Cross-contamination of the aerosols between the hoppers was prevented or minimized by a clear sheet of unexpanded polystyrene covering each vessel. These translucent covers allowed light to reach the algae and larvae as well.

There was never more than one active hopper exposed at any given moment. To make cleaning easier, each hopper may be simply removed off of its table mount. The hoppers are cleaned with a soft sponge and a chlorine disinfection solution after being rinsed with hot, fresh water. Before being filled again, they are carefully cleaned with freshwater. To avoid polluted water from settling or disinfectant from becoming caught in the airline, the airline is left open during the cleaning process. Only individuals who are specifically involved in the production of postlarval Stage IV lobsters should be allowed access to the hatchery. When entering the hatchery, foot baths with chlorine were utilized and all equipment was carefully cleaned in hot freshwater before and after use. Plastics were disinfected with chlorine baths while permeable materials, like nets and brushes, were disinfected with iodine baths. No equipment entered the hatchery without first being cleaned and sanitized. Stage I larvae that had just hatched were quickly removed from the hatching vessel using a cleaned aquarium net, dab dried on tissue paper and weighed. The larvae were then immediately transferred into a 90-liter, aerated conical-shaped hopper that also contained Artemia cysts that had been decapsulated with a chloros solution and incubated for about 12 hours, as well as two species of phytoplankton that had been cultured aseptically.

CHAPTER 10

Ecology of Lobsters

CONTENTS

10.1 INTRODUCTION

Due to their size, relative abundance and propensity to stay in or close to build shelters, the majority of lobster species that live in benthic areas of shallow water are easily accessible for research in the wild. The majority of these species are also adaptable to handling and climatic conditions in aquarium systems used in laboratories. These qualities make them ideal research subjects for both field and lab work. The fact that species of the families Nephropidae (clawed lobsters) and Palinuridae (spiny lobsters) support some of the profitable fisheries in the world is another equally significant consideration. As a result, utilizing techniques developed by fisheries biologists, the population ecology and associated traits of the species involved in these fisheries have occasionally been thoroughly studied. While the demand for both nephropid and palinurid lobsters, as well as the monetary worth of the harvest, are still rising, many lobsters' fisheries' captures per unit effort has significantly decreased. These changes have increased interest in finding out more about the species' habitat needs, population ecology and behavior relevant to fisheries so that the consequences of the fishery can be assessed and forecast as a foundation for efficient management. Additionally, due to these changes, there is a greater interest in creating fisheries for other lobster species and stocks, such as *Panulirus penicillatus* in Micronesia. Much of the knowledge that is currently accessible about the environmental relationships and behavior of nephropid and palinurid lobsters throughout their benthic stages comes from efforts in such fisheries-oriented studies.

The family Scyllaridae (slipper lobsters), the family Synaxidae (coral lobsters), the deep-water species of the Palinuridae and many other palinurids that are not of commercial value or support only tiny fisheries are particularly affected by this in general (Lavalli & Spanier, 2007). These knowledge gaps are regrettable because many of these forms differ from the well-studied species in terms of their larval life cycles, requirements for benthic habitat, and, most likely, in terms of the population dynamics and behavior. Similar to this, our understanding of the ecological traits of palinurid lobsters is based on thorough research of a small number of species, all of which live in shallow water. As a result, although we do have a good understanding of the ecology of lobsters, it is far from complete. For purely practical reasons, fishery biologists who study lobsters typically obtain ecological and behavioral information that directly affects fishing issues that is mostly obtainable from sampling the fishery itself. Such fishery information on abundance and distribution, population size distribution, occurrences of death, birth, growth, mobility and other fishery-related behaviors are beneficial to all those researching lobster in addition to fishery biologists. Among the fundamental ecological and behavioral traits of the species of lobster considered.

10.2 RESEARCH

For obvious reasons, laboratory research has been heavily relied upon to more thoroughly examine the developmental, ecological and behavioral traits of lobster larvae. This method has been utilized quite successfully in research on nephropid, palinurid and scyllarid larvae despite the constraints of such investigations, which are brought on by issues with maintaining a suitable environment and the entrance of other laboratory artifacts. This method has been utilized in a number of studies, including behavioral responses of *Homarus americanus* larvae to hydrostatic pressure and light. Our understanding of the environmental tolerances and behavior of *H. americanus* larvae has improved as a result of laboratory work done with the primary goal of establishing procedures for commercial aquaculture. Our ability to research lobster populations in the benthic environment has also improved. Future research with additional lobster species in deep water will undoubtedly heavily rely on similar methods. Because of their morphology and inability to swim well, phyllosoma larvae are thought to be largely at the mercy of water currents, though there are no direct observations to support this. During the night, the puerulus arrived and settled on the collectors. The puerulus may have used the approaching water currents to help them swim into the location on the evenings when colonization took place.

10.3 LARVA

Phyllosoma larvae exhibit a potent photopositive response to low light. Even under full moonlight, early phyllosoma larvae of *P. cygnus* were localized at the surface at night. Recent research has revealed that on stormy nights, *P. cygnus* early-stage larvae were unable to focus at the surface. The larvae became haphazardly dispersed across the mixed layer as a result of turbulence mixing the higher waters. The latter stages could obtain their chosen depth distribution regardless of the weather since they were powerful enough swimmers to overcome the effects of turbulence. Phyllosoma larvae have been found by several other researchers at far deeper depths. There were more larval densities at 500 and 600 m than at the surface or at 150, 250, or 400 m (Torres et al., 2014). No scyllarids were discovered below

2000 meters, however, some palinurid larvae were found as deep as 3500 meters. Most of the samples from extremely deep water were all late stages. Although it is unknown what these specimens are, it is probable that they are the larvae of deep-water species belonging to some of the more uncommon genera. It has been discovered that the Gulf of Mexico's Panulirus larvae are only known to dive to depths above the thermocline.

Similar to other planktonic crustaceans, light intensity is likely the single most significant factor affecting the larvae's dispersion at different depths. *P. cygnus* phyllosoma larvae migrate vertically every day, rising at night to the water's top and dropping during the day to deeper depths. The dispersion of the larvae in terms of depth was mostly influenced by light. During the day, later stages were typically found at depths of 80–120 m, while freshly hatched larvae were typically located in the upper 60 m of the water column. The higher depth of light penetration in the cleaner marine waters where they are found contributed to the deeper daytime depth distribution of late stages. The later stages only came to the surface on nights that were completely dark and avoided it on the evenings when lunar illumination was greater than 5% of full moonlight. No matter how much moonlight was present, early stages were discovered at the surface at night. Similar patterns have been seen in the settlement of the puerulus of *P. argus*, with the majority of recruitment restricted to the new and first quarter moon phases. Additional research revealed that the puerulus of *P. cygnus* settles on the collectors almost exclusively at night, just before the new moon. The highest concentration occurs when there is a full moon, but it gradually decreases after the moonlight intensity rises above a threshold of around 10% of full moonlight. According to some theories, the presence of puerulus on the collectors during times other than around the new moon is a result of how intense the moonlight is in the region. Despite the extremely low light levels during the last four nights of the last quarter of the moon, settlement does not begin until after the beginning of the dark phase, hence light intensity alone does not adequately represent the pattern of puerulus settlement. It is unclear how the lunar phase affects the puerulus to hinder settlement, just as it is unclear how moonlight can do the same.

No species' level of larval mortality, nor the reasons behind it, is well known. It is likely incidental that phyllosoma larvae occasionally show up in some fish's diet. Phyllosoma larvae of *P. argus* were found in skipjack and blackfin tuna stomachs off the coast of Cuba. However, there have been two instances of what appears to be direct predation. The phyllosoma larvae can move, however, swimming is not one of their best adaptations. It has been observed that the first two phyllosoma stages of *P. japonicas* are fewer adept swimmers than the third and later stages, which feature a feather-like exopod on the third thoracic leg. Developments in the later stages led to further gains in swimming ability. The late-stage phyllosoma swam at a rate of 2-3 cm/sec, whereas the early-stage phyllosoma did so at a rate of roughly one-third of this. In aquariums, first-stage phyllosoma rose vertically at a rate of 0.5 to 0.6 cm/sec. Early Panulirus phyllosoma larvae have been reported to have completed a 5–10 minute swim across a 3.6 m tank in the lab (Bermudes & Ritar, 2008). A thorough understanding of the distribution and abundance of larval clawed lobsters has been limited despite many efforts over an extended period of time due to the inability to harvest them with plankton tows in significant numbers. Although spiny lobster phyllosoma larvae have been collected in greater quantities, the same cannot be said for the puerulus stage. More information is required on this critical phase of the life cycle. Modern methods of collection and analysis are accelerating the solutions, but proper sampling is troubled with logistical difficulties because of the considerable distances the larvae cover throughout development.

10.4 BEHAVIOR

The behavior of many species is inferred from the findings of research on other species due to the dearth of evidence. This kind of generalization, nevertheless, is not always accurate. The absence of an exopodite and the differing feeding habits of phyllosoma larvae of the Scyllaridae and *Jasus* sp. in comparison to the Panulirus larvae, as previously mentioned, raise the possibility of other differences. In fact, many species' rates of development and lengths of larval life are known to vary. The puerulus stage of Jasus and Panulirus has also been shown to react differently to certain collectors. The complete dearth of knowledge on stage-specific mortality rates in the wild makes it impossible to establish any correlations between larval abundance and recruitment to populations, which is another significant area of unfamiliarity. The adaptive significance of a planktonic larval stage in the life cycle, as well as the explanation for the vertical migration behavior, the reason for the distinctive phyllosomal shape and why the larval life is so prolonged in the palinurids and scyllarids, are all interesting topics for research. Although the presence of a planktonic larva affects a wide dispersion, this would seem to be a side advantage rather than the primary cause of protracted larval life. Successful recruiting must therefore be the main goal rather than dispersion. The existence of planktonic larvae among benthic species may represent a generalized adaptation that has been preserved to allow those species to take advantage of the high productivity of the plankton.

The extended larval life and the behavioral modifications connected to this stage in the palinurids and scyllarids are probably techniques to maximize exploitation of the productive plankton population and achieve successful recruitment. Numerous aspects of the ecology, physiology and behavior of larval lobsters can be studied. This could offer some useful information for understanding the ecology of lobster larvae in their natural habitat. The behavior of the larvae is a significant scientific problem. Additional justifications for continuing to investigate this stage of the life history include the animals' economic importance and the possibility of larval recruitment in some species from sources other than the region of release. The ecology of for none of the palinurid species has a sufficient description and frequently even the most fundamental details are missing. Fisheries research has provided some specifics; nevertheless, while these data form the basis of our understanding of many Palinuridae, there are several drawbacks. Economically significant species typically have research that focuses on narrow and particular management-related objectives (Priyambodo et al., 2020)based on a significant natural resource of settling puerulus. These seed lobsters can be captured and on-grown to generate high value, consumption size lobsters.

However, existing fisheries regulation prohibits fishing of these seed lobsters in Indonesia. Nevertheless, seed fishing occurs widely, and the seed are smuggled into Vietnam, where the Vietnamese gain most of the commercial benefit. By removing access to pueruli and grow-out opportunities, these regulations have diminished the capacity of Indonesian lobster farmers to develop a sustainable grow-out industry. Although the new policies were introduced as a management intervention to protect adult lobster stocks, they have had the unintended consequence of preventing development of lobster aquaculture while having no known benefit on abundance of adult lobsters. This study collected data on the puerulus available in Indonesia, species composition, seasonality, and capture methods.\nA census of seed capture determined that 5,243,887 pueruli were fished from two bays in the southeast of Lombok in 2014, representing 64,787 pueruli per km^2. Surveys beyond Lombok determined there was high abundance of seed in areas with similar environmental characteristics, where local communities had also engaged in seed fishing. The locations identified as having substantial puerulus abundance comprised the southern coast of Java, Bali, Lombok and

Sumbawa covering a distance of 1500 km. Puerulus abundance is highest in enclosed bays with distinct currents, relatively high turbidity attributable to terrestrial inflows, and muddy/sandy substrates. Satellite imaging identified all such suitable locations for puerulus settlement, enabling an estimate of total puerulus catch of 103,480,283 per year.

The composition of puerulus fished consisted of two species, Panulirus homarus and P. ornatus. P. homarus was most abundant representing between 63 and 87%, while P. ornatus represented 37 to 13%. Seasonality of catch was unclear, with low catch between December and March and higher catches from April to November. The magnitude of the estimated puerulus resource of Indonesia is >20 times greater than that of Vietnam. If used for aquaculture, the seed available can support an industry generating >12,500 t of market size lobster, and provide social and economic benefit to many thousands of households. To achieve this, fisheries policies could be revised and a regulatory framework established to manage the puerulus fishery and support lobster farming. Tropical marine rock lobster aquaculture is a major opportunity for high-value aquaculture production for the Indonesian economy.”,”container-title”:”Aquaculture”,”DOI”:”10.1016/j.aquaculture.2020.735563”,”journalAbbreviation”:”Aquaculture”,”page”:”735563”,”source”:”ResearchGate”,”title”:”Assessment of the lobster puerulus (Panulirus homarus and Panulirus ornatus, Decapoda: Palinuridae. When it comes to population size-frequency or the start of the breeding season, adequate data are frequently available. Frequently, there is no accurate description of the pristine, unfinished population. The numerous Palinuridae that are not economically exploited are much less well-known. Additionally, funding for fundamental research has been limited, some species are extremely rare and hidden and some only live in deep seas. There are occasionally intriguing hints of unusual behavioral and ecological traits for these comparatively little-studied species. To further understand the family’s basic ecology, especially for the deep-water and/or underutilized species, much future work will be required. Despite these drawbacks, the ecological pattern of the Palinuridae is intriguing and highly varied. They warrant careful investigation because they are one of the most productive and economically exploited benthic decapod groups.

10.5 MECHANISMS

The preference of a species for a particular environment and the behavioral mechanisms connected with this inclination can be characterized as habitat selection. The palinurids utilize a wide variety of habitats and resources. Species that live below 200 meters are currently unresolved (Paulus, 2021). Taxonomically speaking, Palinuridae genera can be loosely associated with latitude, sea depth and temperature based on external morphology. The eight genera have deep-water equatorial (Puerulus, Palinustus, Justitia and Linuparus), shallow-water equatorial (Projasus, Palinurus and Jasus) and high latitude distribution patterns (Panulirus). Although helpful for systematic investigations, such a classification is less successful here due to gaps in our knowledge and differences within genera. Instead, a basic discussion of the palinurids' ecology by habitat, species, age, reproductive status and behavior will be made. The Palinuridae live in a region that stretches from around 45°N to 45°S latitude. In tropical/subtropical waters, there is a wider variety of species, but there are also lower numbers. In temperate waters, species diversity is typically lower but densities are frequently higher. As would be predicted, molting happens more frequently and development rates are accelerated in warmer environments, albeit the maximum adult size might be smaller.

At depths of less than 3 m, mangrove roots, piles, grass beds and grassy undercuts are among the places where the initial benthic stages of shallow-water Palinuridae (*Panulirus argus*) can be found. As the puerulus can be seen swimming freely at night when drawn to the surface by a lantern, settling takes place after metamorphosis from the phyllosome to the puerulus stage. The puerulus is thought to be able to exert great control over its spatial and temporal positioning in the environment to maximize its progress into a suitable, shallow-water. Few investigators have come to the conclusion that the ultimate metamorphosis to the puerulus stage occurs offshore, followed by directed, relatively long onshore migrations or drifts. The ideal environment for the settling puerulus of shallow water palinurids is typically thought to have a fairly shallow depth. On all sides, the seamount quickly descends to considerable depths. Probably preventing any adult immigration to the seamount is its isolation by depth. Adults being present can therefore be thought of as an illustration of a shallow water palinurid settling in rather deep water. The puerulus stage may be able to settle out in both shallow and slightly deeper locations for some shallow-water species. There is evidence that some populations of palinurids are constrained by the juvenile habitat. Recruitment for *P. cygnus* in the waters off Western Australia is characterized by density-dependent mortality in nursery sites, which includes inadequate food availability and limited shelter. This is likely true for many other species as well. When they reach sexual maturity, or just before, shallow water palinurids typically depart their juvenile habitat. Some species, like *P. argus* and *P. cygnus*, may undergo a dramatic mass migration to the adult habitat. This appears to involve *P. cygnus* moving into an adult habitat offshore after synchronized molting into a white stage. For other migratory species, however, long-term photoperiod, temperature, and/or food signals may be significant factors.

10.6 HABITAT

The Palinuridae have a variety of preferred adult habitats, from coastal surf zones with shallow water to deep ocean depths with constant darkness and soft substrates. Along the coast of East Africa, where the occurrence of palinurids has been successfully investigated, there is a reasonably high diversity of lobster species. Five species of Panulirus were divided into several habitat types through research of sympatrio palinurid species off the coast of Africa. These habitat types were categorized by water temperature, turbidity, bottom type, depth, tidal range and geographic area. The most turbulence-tolerant species, *Panulirus homarus rubellus*, was discovered to live in the surf zone and its near surroundings. *P. homarus* can move steadily through the rocks. This species, which is unusual for most Palinuridae, was frequently discovered buried with only its antennae and anterior cephalathorax exposed. It is tolerant of muddy water and suspended sand. Although *Panulirus penicillatus* was discovered in the inshore reef zone, its habitat was less turbulent and sand was not suspended in the air. A variety of habitats, including clear water reef areas and the murky shallow waters of Durban Harbor, were found to be suitable for *Panulirus longipes*. However, it seemed to favor remote, deep reef holes.

It was found that at least one deep-water species has behavioral adaptations for the soft, open bottom that is typical of deep water. *P. delagoae* gathered in open places and noticeably avoided man-made dens even after being placed in holding tanks with appropriate cover. This behavior contrasts with that of shallow water palinurids, which when stressed aggressively seek cover. This survey's findings on other deep-water species did not show much evidence of habitat selection overlap, at least by depth. *Palinurus gilchristi* was discovered in shallower waters, while *Panulirus delagoae* was discovered near the edge of the continental shelf in 180-400 m on a high organic content mud substrate (Mahdi Abkener, 2020). Spiny lobster species that live in shallow water and on hard substrates are significant commercially, but they do not entirely represent the Palinuridae family. It's possible that the absence of numerous predators and the stable environment at large depths reduce the need for hard substrate refuge while still supplying a suitable habitat. But it's important to question our presumption that these deep-water animals only live in open, soft substrate habitats. Our data primarily come from commercial fishing trawlers, however, correlations between deep-sea species and soft mud substrates may be biased since fishermen tend to avoid rocky substrates in an effort to reduce expensive equipment damage.

In the Indo-West Pacific, *Panulirus* spp. species that live on reefs have been classified by habitat preference. For many of the palinurids, the living quarters or den is the habitat's most distinctive element. The Palinuridae don't often build dens; instead, they opportunistically select them from the surrounding habitat (DeVantier, Turak, & Szava-Kovats, 2020). The majority of Palinuridae are typically thought to have limited ability to burrow. *P. argus* cannot burrow and must rely on any available shelter. For the majority of palinurids, there isn't much explicit evidence of burrowing or tunnel maintenance in the wild. This presumption might alter if more in situ observations can be made because of the usage of scuba. Sand and small stones that block the den entrance are pushed by captive lobsters (*P. argus*) to enlarge and otherwise alter their dens. The lobsters either plowed the sand forward by digging the anterior pereiopods into the substrate or they trapped a tiny rock between the basal segments of the antennae and moved it forward by walking. In experimental tanks, trail-flap-induced sand movement was also noticed in this species' juveniles, as well as in another species related to defense. There is a strong probability that some palinurids have the ability to make small den modifications outside of the lab.

10.7 DISTRIBUTION

The distribution of lobsters among potential den sites is never random) and the selected dens may be distinctive. No physical characteristics, such as aperture size, height, or volume, were highly linked with the den sites. In contrast to the conventional two-opening construction of *P. homarus*, *Panulirus argus* dens have just one opening. Dens, however, were frequently identified by their hard substrates and partially constrained apertures. Other unmeasured factors, including availability of feeding grounds, current flow and water quality, may have also played a vital role in the choice of den. Although *P. homarus rubellus* is not involved in the construction of such dens, it was noted that *P. homarus rubellus* frequently occupied dens that were tunnel-shaped with openings at both ends. Palinuridae individuals are social both within their own species and frequently between them. It is occasionally discovered that *Panulirus argus* and *P. gutattus* share a den, though usually not in close proximity to one another. In contrast to *Panulirus argus*, which is typically found on the den floor, *Panulirus gutattus* has frequently been seen clinging upside down to the den roof or withdrawing into tiny cracks in the roof and walls. In the Caribbean, *Panulirus argus* can occasionally be found living alongside the spotted moray (*Gymnothorax moringa*) (Figure 10.1) or the green moray (*Gymnothorax funebris*) (Figure 10.2). It became clear that even a small moray had the potential to make a rapid meal of an average-sized *P. argus*, but presumably often chooses not to use a tethered lobster to annoy a spotted moray that was residing with a number of *P. argus* in a huge den.

Figure 10.1. *Gymnothorax moringa.*

Source:https://www.fishbase.se/images/species/Gymor_u0.jpg

Figure 10.2. *Gymnothorax funebris.*

Source:https://www.fishbase.in/images/species/Gyfun_u0.jpg

10.8 SOUNDS

The stridulatory sounds made by the inmates of the lobster dens appeared to entice *Lycodentus undulatus* to them. The benefit of this relationship for the local lobsters is evident and it might also attract octopuses as prey for the moray. Palinurid reproductive biology is highly diverse and environmental factors undoubtedly have an impact. Even for just one species like *P. argus*, the time of mating and egg production has been described as seasonal, numerous, or basically continuous in various parts of its range. *P. argus* only has one significant mating season, which occurs in the spring. This high rate of reproduction indicated two peaks or perhaps one considerably prolonged reproductive period every year for *P. argus* at this location, as it persisted into the early fall, which is typically considered to be nonreproductive for this region. *P. argus* has two neighboring breeding seasons in Brazil and Cuba. Four peaks of reproductive activity were noted in the Caribbean. The variations in this species' reported breeding seasons are likely the result of variations in the local environment, particularly in terms of temperature but also in terms of nutrition and photoperiod. These variations most likely result from the timing of molting being different for large and small females, which for this species closely coincides with mating.

Diverse data collection techniques are partly to blame for differences in data for adjacent populations of the same species. It has been demonstrated that females and newly molted lobsters are solitary and less likely to fall into traps. In order to quantify the bias caused by trap sampling, it is essential to supplement data collected using traps with observations and scuba-based individual captures. Trap sampling can move forward with accurate interpretation once this has been completed for a particular location. It is likely that a variety of parameters, including temperature, food quality and availability, substrate suitability, population density, water chemistry and others, combine to affect the number of spawnings, molts and overall fertility for many palinurid species. Many tropical species may be able to breed continuously under the right circumstances. *P. cygnus* reproduction in natural populations is distinctly seasonal. Although a direct link between temperature and reproduction has not been established, it is likely that temperature has a significant role in determining when a palinurid species will reproduce. But the effects of density on growth rate in *P. cygnus* mask the parameters governing size at first breeding. The frequency of repeated spawnings in the wild is governed by a variety of circumstances and may serve as an indicator of the general ecological suitability of a species. In certain palinurids, molting and sexual activity occur simultaneously.

According to some research, male *J. lalandii* will become more active when females are in the premolt stage. *Panulirus argus* conducts external fertilization, therefore mating is not always required when the female is in the soft state. However, it was found that females do mate soon after molting, though most likely after the carapace hardens. It is challenging to identify or measure size at sexual maturity for a population of a particular species. Size is most likely equally variable over the range of a given species as spawning frequency in terms of its sensitivity to environmental factors. The average size of reproductive females appears to not vary significantly across numerous Palinuridae species in a given region. The ability to carry eggs is obviously correlated with female size. According to estimates, *P. argus* lays and carries about 800 eggs on average per gram of body weight, with estimates ranging from 600 to 1200 eggs per gram. Therefore, for evaluating overall population fecundity, a method that takes into account female size-frequency structure, as well as other relevant factors, should be used. When defining legal size limitations, we must examine all aspects that affect a population's capacity to contribute to the fishery's future recruitment, not only the small, sexually active females. This is necessary to preserve a population's overall capacity to do so. The creation of more practical policies may be made possible by the simultaneous treatment of several variables. However, it must be emphasized that data for such therapy must be gathered from populations that are not subject to intensive fishing. If not, the analysis needs to take into account the skewing bias in favor of petite females caused by the commercial cropping of larger individuals. It is interesting to note that very large females were proportionately less frequently observed to carry eggs than intermediate-sized females. It is unknown whether this is caused by older females molting less frequently or by sexual senility that affects huge females.

The idea of relative reproductive relevance ought to guide our cropping decisions, as well as legal size restrictions and could be helpful in creating future management strategies for lobsters. Some Palinurid populations, nevertheless, might be limited by habitat rather than recruitment. Such circumstances would make this strategy less effective. However, the majority of populations do not know their limiting variables and prudent management should compel us to believe that total egg production is a significant role. Many local protective lobster management policies need to be evaluated in light of this. It can be problematic to predict how long an egg will carry in the field. The dependability of laboratory data may be harmed by captive conditions, which frequently do not precisely reflect field circumstances.

In order to evaluate the amount of time spent in each reproductive state in situ, a method has been proposed that looks at the percentage of fouled females i.e., with external growth on the carapace and algae observed for each reproductive state and the frequency of each reproductive state within the sample population.

There are two types of fishing pressure on the Palinuridae: recreational fishing and commercial fishing, which is significant. The impacts of strong fishing pressure on unfinished lobster populations have not received much attention. Typically, research doesn't start until the population has manifested the harmful impacts of overfishing and assumptions are frequently limited to pre-exploited conditions. The fishery's behavioral impacts were less obvious after 16 months, but the overall density remained significantly lower than the initial level. Such research on protected populations is both uncommon and valuable. Animal behavior and the environment are closely related. The most evident and possibly well-researched stimuli for the palinurids among the many potential influencing factors are temperature and light. Variations in the ambient temperature are frequently correlated with population distribution and/or behavior. Reproductive immigration and emigration from deep reef areas caused these fluctuations. According to Sheard's hypothesis, Palinuridae species that naturally undergo significant annual temperature swings are more likely to display ecological correlates with temperature peaks and lows. These palinurids reproduce when the temperature is higher in the spring and their eggs hatch when the temperature is highest in the summer. The reproductive cycle of *P. cygnus*, in contrast, does not correspond as well with temperature in Western Australia. Here, mating takes place well before the lowest winter temperature is achieved and the eggs hatch well before the highest summer temperature is reached. Like other Crustacea, palinurids are affected by temperature when it comes to molting and growth. *P. interruptus* grown in thermal effluent from a power station molted more regularly.

In areas where the temperature fluctuates significantly and relatively regularly, it is logical to infer that it affects when animals reproduce, molt, migrate their populations, etc. Although a connection with a particularly spectacular behavior pattern has been documented in laboratory and field experiments, the full impact of temperature on the biology of Palinurids has not yet been determined. It's interesting to note that when the temperature drops, both the length of activity and the nature of the behavior portrayed change. Most likely, temperature directly influences the behavior and activity patterns of other Palinuridae, particularly those migratory species that are naturally subjected to significant temperature changes. There are a variety of other environmental factors affecting the ecology of the Palinuridae, but they are less well known. The biomass of the related benthic algae is often lower than that of the algae found in the bedrock base with rock habitat due to the larger depth distribution of the sand-rock substrate. Additionally, although ascidians are similarly common in both habitats, the abundance of giant anemones is relatively lower. The brachiopod Terebratulina is widespread below depths of 20 meters and there are many sponges that are clinging to rocks and stones. As compared to the bedrock substrate, the mollusks Mytilus and Modiolus are rare, but rock crabs and green sea urchins are extremely common.

It is impossible to find macrobenthic algae in the offshore sand-rock habitat. On the exposed surfaces of rocks and boulders, there are extremely high populations of anemones, while sponges are in short supply. There are numerous polychaetes that live in tubes and hermit crabs that lack shells. For lobsters inshore but not offshore, the bedrock base with rock and boulder overlay is a fairly typical habitat type. The bedrock substrate typically reaches a depth of 15 to 45 meters from low tide. It doesn't happen in the offshore lobster areas, with very few exceptions. Crevices and rock overhangs provide protection for lobsters. Lobsters may take refuge inside the rock pile if there are enough rock/boulder accumulations there. Generally speaking, this substrate makes it impossible to burrow. To depths of 42 m, green, red and brown macrobenthic algae offer varying degrees of shelter. Up to 90% of the visible bedrock is covered in calcified, encrusting coralline algae. Most encrusting or adhering creatures are brachiopods, mollusks, large anemones, ascidians, sponges and mollusks. The most prevalent echinoderms are the green sea urchin and starfish. The fish that are most prevalent in the bedrock-rock habitat are cunner, tautog, sculpin, sea raven and redfish. Another characteristic habitat for lobsters is the mud foundation with burrows.

Mud anemones, shrimp, rock and Jonah crabs and various fish species, such as squirrel and white hakes, redfish, witch flounder, four-bearded rockling, ocean pout and goosefish, are the macrofauna that is most frequently found in the mud-burrow habitat with the lobster. Due to the lack of exposed hard surfaces, attached or encrusting organisms and macrobenthic algae are typically absent from this habitat type. The upper slope and outer shelf both have a clay base with depressions and burrows, which is typical of the inshore mud-burrow environment. Particularly in the region of underwater canyons, a significant portion of the outer shelf and upper slope of New England and the Mid-Atlantic Bight have a surface substrate of Pleistocene clay that is largely free of overlying soft sediments. These clay environments have a relief range of 0 to 5°. The surface of the exposed clay is predominantly eroded by lobsters in the form of bowl-shaped depressions. These depressions typically have a symmetrical pattern and extend up to 1.5 meters into the clay substrate. They frequently have a solid object, such as a boulder, rock, coil of steel towing rope, fish net, or fuel oil drum, in the center of the depression. However, current movement and bottom scouring make it easier to remove silt from the area around a solid item.

10.9 MACROFAUNA

The main macrofauna associated with this clay habitat are mud anemones, pandalid shrimp, Jonah crabs, galatheid crabs and black-bellied rosefish. When there are depressions where lobsters are present, occasionally squirrel hakes and Gulf Stream flounder are seen there or very nearby. Red crabs, long-finned hakes, four-spot flounder, witch flounder and goosefish are regularly seen in the clay-depression environment at the deeper depths of the lobster's distribution. On the outer shelf and upper slope, the mud-clay base with anemones constitutes a relatively frequent habitat for lobsters. Typically, the mud-clay substrate includes a thin silt layer that can be up to 3 cm thick. Inshore lobster grounds rarely exhibit an ecosystem like this. Mud anemone can accumulate in some offshore regions at densities of three or four per square meter. At the base of the tube, a relatively shallow depression is frequently created from excavation or current erosion. These craters provide protection for lobsters that are typically in the 50–80 mm CL size range. Compared to the depressions described for the clay basis with depression habitat, this is far smaller in size. Jonah crabs, bathynectes crabs, galatheid crabs, red crabs, pandalid shrimp, squirrel hake, black bellied rosefish and redfish make up the majority of the macrofauna that is typically found in the clay-anemone ecosystem. Anemone tubes frequently have unidentified species adhering to them from the substrate level up to 12–14 cm above the mud–clay foundation. The submerged canyon clay wall with burrows, which is predominantly made of Pleistocene clay, is the most spectacular type of lobster habitat-substrate found offshore. This clay substrate has been heavily dug by lobsters and other macro-benthic fauna.

Stage IV lobsters that have just moved in are particularly skilled at quickly tunneling into soft soils behind heavy things like stones or shells. Studies in aquariums have shown that the majority or all of the newly established stage

IV larvae dug tunnels and sought refuge there before molting into stage V. These recently created tunnels can be U-shaped or have two openings that are 180 degrees apart on opposite stone edges ("Burrowing - an Overview | ScienceDirect Topics," n.d.). Adult lobsters exhibit the same digging techniques. Depending on the type of substrates, such as mud, sand, gravel, or shell pieces, these approaches are applied to varying degrees. For weeks or months at a period, lobsters may hibernate in a mud burrow throughout the winter and seal up the entrance with a partition of sediment and debris. Typically, the entrance is surrounded by a wall of excavated bottom material, which is most obvious in front of recently constructed burrows. It has been established that pollutants such as heavy metals, pesticides, pulp mill effluents, petroleum hydrocarbons and others have a negative impact on lobster physiology and behavior. Additionally, earlier acclimation or physiological conditioning at sublethal levels of the element being evaluated dramatically increased tolerance to extremes of these factors, either alone or in combination. Pre-adult lobsters can withstand temperatures between 1.8° to 30.5°C at the ideal oxygen concentration and salinity, given reasonable acclimatization, such as can be expected within natural changes in the temperature cycle. Positive correlations were found between acclimation temperature and tolerance to both high and low temperatures. The top lethal temperature range was significantly influenced by temperature, salinity and oxygen acclimation. The range increased as acclimation temperature increased, but decreased as salinity and oxygen levels decreased. Despite their level of adaptation to oxygen, lobsters are surprisingly tolerant of low oxygen levels. Temperature and salinity acclimation were extremely relevant variables but oxygen concentration did not significantly affect fatal oxygen concentration.

CHAPTER 11

Behavior of Lobsters

CONTENTS

11.1 INTRODUCTION

Spiny lobsters typically dwell in shelters in rocky or coral reef settings, carry the eggs for just a brief period and have a long larval life. Only a few adult individuals found among coral boulders and a few larval specimens discovered in plankton nets have allowed scientists to learn anything about coral lobsters. One may almost always expect to find some kind of lobster in a marine area. Some have been trawled up to 3000 meters beneath the surface of the ocean. Others can be picked up by hand at low tide and are found living in holes on small tidal reefs. The cold temperate continental shelf region of the North Atlantic is dominated by two genera of clawed lobsters: Homarus and Nephrops (Stein et al., 1975). However, certain Zoogeography sections or depth zones are typically dominated by various groups of lobsters. On the western North Atlantic's continental shelf, *Homarus americanus* is present, whereas *H. gammarus* and *N. norvegicus* are found in the eastern North Atlantic. In addition, *H. gammarus* inhabits shallower, rocky locations as opposed to Nephrops, which prefers deeper waters and muddy bottoms. Slipper lobsters are commonly found on soft substrates like *Ibacus peroni* and *Thenus orientalis*, both of which are commonly collected during commercial shrimp trawling. However, a number of species, including Parribacus and Scyllarides species, can be found on stony bottoms. The flanged laterocarapacial surfaces and flattened antennal appendages, which are typically assumed to be employed for burial in soft substrates, of *Scyllarides nodifer* are a feature that provides protection by hiding the animal on the hard substrates on which it is collected. The few *Palinurellus gundlachi gundlachi* specimens that have been obtained, one of the two subspecies of synaxids, have all come from coral rubble sites in shallow water.

The clawed and spiny lobsters off the coast of Durban, South Africa, are the best examples of zonation by depth. The clawed lobster *Nephropsis steward* is the only species that can be found at depths greater than 300 fathoms. *N. steward* and *Metaneprops andamanicus* coexist between 300 and 230 fathoms, although *M. andamanicus* predominates on the ledge between 220 and 230 fathoms (Chang et al., 2020). Spiny lobsters of the genus *Panulirus*, particularly *P. homarus* rubellus, predominate in the shallowest waters from coastal reefs and offshore shoals to about 10 fathoms on the continental shelf. Spiny lobsters of the genus Palinurus (a spiny lobster) are the dominant crustacean trawled at shallower depths between 60 and 80 fathoms. There is vertical zonation of the nephropid lobster subfamilies in the Atlantic. The commercial clawed lobsters are members of the Nephropinae, which inhabits the shallowest seas between 0 to 885 meters (Kampouris et al., 2021). Thymopinae and Neophoberinae, less well-known subfamilies, are found deeper than 200 m and can be found down to at least 3000 m. Temperature is widely regarded as a major ecological characteristic. Although several genera like *Jasus*, *Palinurus*, *Linuparus*, *Justitia* and *Palinustus* have a similar preference for a temperature of 10°-20°C, they rarely occupy geographical regions that correspond in latitude and depth. Species of the genus *Panulirus* almost exclusively occupy the warmest low latitude regions of the continental shelf. *Projasus* and *Puerulus* species, which belong to the most primitive genera, predominate in the subpolar temperate zone. Once more, each genus is geographically distinct from the others.

11.2 FACTORS

Environmental factors that are biological and physical have an impact on species dispersion. It is necessary for physiological tolerances to environments to allow for both reproduction and survival. Larval recruitment to the juvenile stock must be assured by the water currents. Local species interactions including competition and predation, as well as individual movements and habitat preference, limit abundance. Limiting variables frequently include the physical conditions of temperature, salinity and oxygen. Although in-depth experimental assessments of physiological tolerances to these parameter extremes have not been conducted for the majority of lobster species, ranges of reported fluctuations can be provided. The cool waters are the only places where nephropid lobsters can be found. Although larval stages of *H. americanus* tend to be restricted to temperatures between 10° and 25°C, adults can endure water temperatures between 1° and 28°C. Nephrops' physiological tolerance is unknown, but the species has been observed in temperatures ranging from 7° to 15°C, suggesting that the real range is undoubtedly wider (Qadri et al., 2007). For the juvenile and adult populations of lobsters, salinity does not appear to be a significant limiting factor, however, there are signs that larval Homarus cannot tolerate exposure to salinities below 17 percent O2 and will avoid water that is 21.4 percent O2 or less.

Temperature, salinity and oxygen tolerance depend on one another. Living close to one's tolerance limit will make it harder to withstand a significant shift in another element. Lobsters choose among the numerous settings that are available to them depending on their tolerance for the aforementioned characteristics. Overall, an animal's physiological tolerances, the availability of food, a suitable substrate, predators and competitors all have a role in a habitat's suitability. The substrates used by lobsters range from rocky regions to reefs of limestone or coral, to deep, soft mud suited for tunnels. All of the substrates appear to have the same feature of offering a tunnel or crevice where the lobster can hide from predators, harsh light and other unfavorable conditions. While spiny lobsters are typically located in communal dens, clawed lobsters typically find individual protection. Another distinction is that in contrast to juvenile and adult clawed lobsters, which frequently coexist, young spiny lobsters are typically found in shallow protected regions like lagoons in sponge and weed beds. Young animals like *P. cygnus* remain in these nursery sites until they are 3–5 years old, at which point they migrate offshore, mature and reproduce. The installation of man-made cement shelters in a sandy area generally unused by Homarus provides proof of the significance of shelter in habitat selection. Within a week, 60% of the shelters were inhabited and a month after the placement of the shelters, the area had more lobsters than shelters.

11.3 COLORS

The green, white, black, yellow and pink hues of the spiny lobster, *P. versicolor*, are patterned in lines, bands and stripes. Many kinds of spiny lobster have vivid markings on the dorsal side of the abdomen and back. The dorsal and lateral patterns on the abdomen and back of spiny lobsters are only visible when they emerge from their shelters. They typically migrate more or less permanently to another place or temporarily leave their shelter to forage when out at night. The importance of a dorsal, distinctive belly pattern is probably unrelated to feeding activity as feeding motions are more frequent on moonless nights when light intensities are low. However, one species of *Panulirus* is known to migrate in large groups over the course of several days and nights. The two parallel rows of huge white spots spanning the length of the abdomen are the most distinguishing feature of each of these three species' dorsal color patterns. Tactile contact with the inner rami of the antennae keeps each rock lobster in the line of sight, but the initial approach from behind a moving object requires visual orientation.

Because heat denatures the protein to which the colors are attached, all lobsters turn red when cooked. Both nephropid and palinurid lobsters have been found to be able to emit low-frequency sound. Specific sound-producing mechanisms in *Homarus americanus* and six of the eight genera of the Palinuridae have been discovered morphologically. The Palinuridae's stridentes group of genera, which includes *Panulirus*, *Palinurus*, *Justitia*, *Palinustus*, *Linuparus* and *Puerulus*, have a stridulating apparatus made of an expanded ribbed pad on the base of the second antennae, called the plectrum. The second antenna's extended base and the ability to shift the antenna along the constrained vertical antennal plate are not present in the silent group of palinurids. In this group, a stable articulation exists between the anterior basal point and the antennular plate. The plectrum has a chamber underneath it that reverberates, amplifying the sound. With increasing size, the plectrum's number of ridges also increases. *P. cygnus* develops the file and plectrum during the puerulus stage, but sound is not produced until the postpuerulus phases. Numerous species have been quantitatively examined. It is relatively consistent across all species and has frequencies between 85 Hz and at least 11–12 kHz. The peaks of energy at particular frequencies may vary slightly. *P. argus* and *P. guttatus* differ from one another. The length of the rasp varies depending on the species and the size of the individual. The stridulation for *P. guttatus* lasts 53 msec on average, compared to 92 msec for *P. argus*.

The typical crustacean compound eye found in lobsters is composed of thousands of small, square lens-retina units, or ommatidia. An outer corneal lens, a crystalline cone and a rhabdome that is encircled by retinula cells make up each ommatidium. A sleeve of pigment cells that migrate proximally and distally in response to light intensity surrounds the seven retinula cells. Superposition eyes, or those that are only exposed to dim light levels, are often seen in arthropods that are active at night or in the deep water, such as lobsters. The crystalline style, an extended process that separates the crystalline cone from the rhabdome, is one feature of the superposition eye that results in a clean zone between the optical elements and the retina. Arthropods that typically survive in high light intensities have an opposing eye, in contrast. There is no distinct zone in these arthropods, but the retinula cells stretch from the basement membrane to the crystalline cone and each ommatidium is believed to function independently in the collection of light. The relationship between the anatomy of the eye and many vision-related aspects in larval and adult *H. gammarus* lobsters has been investigated. The crystalline cone secretes, rhabdomeres develop, distal pigment cells sprout between the cones and the inner ends of the cone cells enlarge to produce long cone stalks as the eye grows and develops between the larval and adult phases. Electroretinograms produced using the larva and adult are frequently comparable. These investigations made use of a white light stimulus lasting 40 msec. There was an initial, quick negative response at the corneal surface in the larva that was not observed in the adult. The electroretinograms of the larva showed both a fast and a slow negative component. The delayed negative response completely vanished when the electrode was put right beneath the corneal surface, while a tiny penetration caused the quick positive wave to reverse. Only surface records of a fast negative wave were detected in *H. gammarus* larvae. The logarithm of the stimulating energy directly affects the amplitude of the fast potential of the *H. gammarus* eye.

11.4 SENSES

Although the ability to perceive images is presumably conceivable, the level of visual acuity is unknown. The complex eyes of lobsters are excellent at detecting movement. The rhodopsin-based visual system of lobsters has a maximal sensitivity at 500 nm and exhibits little response to light in the red end of the spectrum (Frank & Case, 1988). Although it is unknown if lobsters can distinguish between hues, color vision seems to be common in decapods and there is no reason to doubt that lobsters have it as well. The lobster eye, like those of other Crustacea, appears to be morphologically well-suited for the perception and interpretation of polarized light, but no adequate behavioral or electrophysiological studies have been conducted on lobsters to support this claim. When the eye is fully light-adapted, the screening effect of the pigment layers in the eye partially protects the rhabdomes, but in at least one lobster species, *N. norvegicus*, exposure to 2.5 hours of intense light, as would occur when the animal is brought to the deck of a fishing vessel, bleaches the visual pigment and breaks down the structure of the rhabdomes. The permanence of the blindness resulting from this is unknown. In comparison to Nephrops, other lobsters like Homarus and several of the commercially significant Palinuridae are found in shallower waters, therefore exposure to intense light may not be as harmful to them.

The chemoreceptive capacities of lobsters have been extensively studied in behavioral investigations, ranging from a fisherman's experimenting with various types of bait to the presentation of specific amino acids in well-controlled laboratory tests. All evidence suggests that lobsters are capable of odor discrimination and that scents made up of many compounds are the most stimulating. In addition to their obvious and highly developed visual and chemical capabilities, lobsters also have a highly developed sense of touch and can learn a lot about their environment through this sense. Mechanoreceptors also convey information about limb position, vibration and gravity. The majority of these sensations are carried out by the microscopic cuticle extensions known as cuticular hair organs, which are innervated by one or two bipolar nerve cells. The entire surface of lobsters is covered in tactile sensors. They come in a variety of shapes, but they typically have a peg or hair-like shape and occasionally resemble fans. Some seem to insert through a movable socket into the cuticle. Some are positioned in blatantly asymmetric cuticle depressions and only react to movement along the axis of asymmetry. The ability of lobsters to sense tactile stimuli, such as water currents and to determine the direction of the stimulation, thus, seems to be a given. Low-frequency vibrations may also be detectable by these receptors.

11.5 SEGMENTS

The statocyst in the basal segments of the first antennae allows lobsters to retain their orientation with respect to gravity and detect acceleration. The cuticular surface of the statocyst is folded inward to create a nearly fully enclosed cavity that is lined with mechanosensory hairs. These touch hairs on the lobster are innervated by a single neuron as opposed to the two that other regions of the lobster have. A statolith made of many sand grains that have been bonded together is found inside the hollow that is resting on some of the hairs. When an animal tilts in relation to gravity, the statolith shifts on the tactile hairs, signaling a change to the central nervous system, which then starts corrective leg motions. By bending in reaction to fluid movement inside the statocyst, thread hairs that are not in contact with the statolith are able to sense acceleration. For locomotion and posture maintenance, awareness of limb position and movement feedback is crucial. Numerous internal proprioceptors keep track of postural changes. For instance, chordotonal organs are located at the articulation between joints, while muscle receptor organs (MRO) are located in the muscles of the thorax and belly. Muscle contraction causes a gradual stimulation of the abdominal MROs, which ostensibly detect motions. Mechanosensory neurons embedded in strands of elastic tissue that cross the joints of the walking legs, antennae and maxillipeds constitute the chordotonal organs. The elastic strand is bent by the movement of the joint and this causes the neuron to fire. When the joint is fully extended or fully flexed, some of the cells discharge continuously, whereas others only react to flexion or extension in a unidirectional way.

Crustacean somatic muscle fibers are multinucleated and striated, similar to those of vertebrates. They typically extend from the insertion on an apodeme of the following segment to the origin on the cuticle of the previous segment. The two most common forms of muscle fibers are. The fast fibers, which have numerous sarcomeres, provide quick, easily-fatigable contractions, such as those of the abdominal muscle during a tail-flip escape response. In the closer muscle of Homarus' crusher claw, long sarcomere fibers, known as slows, create slow, forceful contractions that can be sustained for a long period. One to eight axons, up to six of which may be excitatory and one to two inhibitory, innervate the muscles. Slow fiber excitatory axons often have modest diameters and exhibit little fatigue during prolonged nerve stimulation. On the other hand, the axons that innervate the fast fibers tend to be larger in diameter, tire out quickly when stimulated and have a lot of synaptic vesicles. The function of a muscle depends on the characteristics of its fibers, axons and synapses since a muscle can consist primarily of slow fibers, largely fast fibers, or a combination of both. L-glutamate appears to be the transmitter at the synapse of excitatory axons and amino butyric acid is the transmitter at inhibitory synapses (GABA). The supraesophageal ganglion, an anterior cerebral ganglion that is located above the esophagus and is connected to the thoracic and abdominal nerve trunk by two circumesophageal commissures, makes up the lobster's central nervous system. The chain of ganglia in the nerve may be anteriorly concentrated in a big fused subesophageal ganglion, as in *Jasus*, or anteriorly elongated, as in Homarus. The primary association center for the entire body is the supraesophageal ganglion, or brain. It gets neural input from the head's sensory organs and ascending interneurons of the ventral nerve cord. From there, interneurons branch out to the ventral cord's ganglia, which regulate muscle movements. The more sophisticated behavioral patterns are started by the brain, which then adjusts them through learning. It is divided into three parts. The protocerebrum, which is the area closest to the eyes and the frontal lobes, appears to be primarily responsible for beginning behavior. Olfaction is connected to the middle brain's deuterocerebrum because it gets sensory information from the antennae.

Through five anterior and two posterior arteries, blood leaves the heart. The artery system in *P. interruptus* is typical. The cor frontale is formed by the median aorta, also known as the ophthalmic artery, which flows anteriorly and somewhat enlarges over the anterior part of the stomach. This structure functions as an accessory heart, enhancing blood supply to the brain, eyes and antennules and it is connected to two strands of somatic muscle that contract rhythmically. The majority of the other thoracic organs and muscles are supplied by antennal arteries, which leave the heart close to the ophthalmic artery. From the anterior ventral side of the heart, paired hepatic arteries travel to the hepatopancreas. The descending artery and the posterior abdominal artery are formed by the posterior division of the thin-walled bulbus arteriosus. The ventral abdominal artery, which provides the abdomen muscles and some pleopods and the sternal artery, which branches off to supply the walking legs, mouthparts and in the Nephropidae, the chelipeds, are the two divisions of the descending artery. A large portion of the abdominal musculature, appendages and gut are supplied by the posterior abdominal artery. The gills, the stomach and to a lesser extent the antennal or green glands are the primary excretion organs for nitrogenous wastes. The same mechanisms are used to balance electrolytes and water. The two antennal glands, which are positioned anterior to the ventral part of the cardiac stomach, open through large papillae on the side of the first antenna's basal segment that faces downward. They consist of a ventral glandular region and a dorsal, thin-walled bladder.

11.6 AMMONIA

Ammonia is the main excretory substance produced by lobsters. Ammonia accounts for 72% of the total nitrogen excreted in *Jasus edwardsii*. *Panulirus cygnus* has varying degrees of ionic regulation ability and is an osmoconformer within a constrained range of salinities. The bulk ions, sodium and chloride, which are accumulated and decreased, respectively, but change immediately with external salinity, appear to be responsible for the poikilo osmotic properties. Within the salinity range of 25-45 percent, this is enough to prevent fatal dehydration. There is a tiny initial rise in blood volume in a hyposaline medium, which is followed by an increase in total volume that is tolerable. The extra salt is subsequently expelled through the gills until the balance is reached. The appendages' peripheral muscle may act as a salt pool. It's possible that water that has entered through endosmosis is eliminated through an overshoot mechanism, in which the tissues pump out a little bit more ions than are required for equilibrium, causing the water to then be lost through exosmosis to return blood and other tissue volumes to normal. Ions are controlled separately; calcium is stored while excesses are eliminated through the gut. While sodium, potassium and possibly chloride are mostly controlled by the gills, magnesium is expelled slowly through the antennal glands. Additionally, nonprotein amino acids made up about 40% of the osmotically active compounds in *P. cygnus* muscle. The primary organ for removing oxygen from the water in lobsters is their gills.

The swimmerets have a limited respiratory function, using about 3% of the total oxygen breathed. The maxillipeds and pereiopods project the gills into the branchial cavity. They resemble a bottle brush because they are of the trichobranch type and have many tubules on every side of the shaft. The difference between the three types of gills is determined by the location of implantation.

The coxa serves as the origin of the podobranch gills, the arthrodial membrane at the base of the limb serves as the origin of the arthrobranchs and the surrounding body wall serves as the origin of the pleurobranchs. Homarus kept in water with a high 02 level often exhibit unilateral scaphognathite activity and irregular heartbeats. Additionally, this suggests that the interplay between respiratory and cardiac motor neurons within the central nervous system is what causes the strong link between respiratory activity and heart rate. The heart and scaphognathite rates increase and beat continuously when the oxygen concentration falls to 30–50 mm Hg. In both well-aerated water and hypoxic environments, the oxygen intake is unchanged. Scaphognathite beat rates drastically decrease and become irregular in water that is extremely lacking in oxygen and the consumption of oxygen is significantly decreased. Thus, lobsters are able to control their oxygen consumption throughout a broad range of 02 concentrations in their environment.

The lobster's foregut, midgut and hindgut are its three distinct digestive system areas. The foregut is rather straightforward and straight. Uncalcified chitin, which is shed during molting, lines the fore- and hindgut. The mouth, esophagus and a two-chambered stomach make up the foregut area. The esophagus is a narrow, muscular conduit that connects the mouth to the stomach's cardiac region. Chitin lines the walls, which are made of columnar epithelium. Around the labrum and at the front part of the esophagus, there are a lot of tegumental glands. These release a mucopolysaccharide that functions as a lubricant, it seems. Between the esophagus and the cardiac stomach lies a trilobed valve. An anterior cardiac region and a posterior pyloric portion make up the stomach. The thick chitinous lining of the foregut's heart region is calcified in some places to create the gastric mill. Three big opposing teeth, one dorsal and two lateral, make up the mill, a grinding device that triturates and separates the food particles. The grinding movement is brought on by strong muscles on the stomach wall. Two calcareous plates, or gastroliths, are seen in the cardiac stomach of nephropid lobsters. The spiny lobsters do not contain gastroliths. There is just one dorsal caecum and two ventrolateral ampullary filters in the pyloric portion of the stomach. Here, the food is sorted and strained before being sent back to the intestine, the hepatopancreas, or the gastric mill.

11.7 GLANDS

On either side of the stomach and anterior part of the midgut, there are two pairs of digestive glands called the hepatopancreas, which are yellow-green in appearance. The hepatopancreas has openings into the area where the pyloric stomach and the midgut meet. The gastric mill's movements thoroughly combine food particles with enzymes generated by the hepatopancreas in addition to masticating the food. Although the stomach secretions are acidic, it is unknown what keeps the pH low. The hepatopancreas, the only region of the alimentary canal other than the midgut that is not walled with chitin, is where absorption into the bloodstream primarily occurs. Circular and longitudinal muscular tissue surrounds a columnar epithelium that makes up the midgut. The debris is propelled along and into the hindgut by the peristaltic movements of the midgut. The longitudinal musculature of the hindgut is similar to that of the midgut and is fairly short. Contrary to those of the midgut, the circular muscles of the hindgut are thick. The anterior region of the hindgut is rich in tegumental glands, which are marginally smaller than those in the anterior portion of the esophagus. The anus, which opens as a longitudinal slit on the ventral surface of the telson and is managed by external radial muscles, is where the hindgut exits.

Indicators of nutritional status in lobsters include serum protein levels. In *H. americanus*, hunger results in a drop in serum protein concentration. The concentration of blood protein varies due to dilution in *P. cygnus*, but the total amount is stable. Although they have a reputation for being omnivores and scavengers, lobsters really appear to be rather selective in what they consume, as evidenced by the fact that the proportion of food items in the stomach differs from their relative abundance in the lobster habitat. For instance, compared to their relative densities in the field, the proportion of crabs in the diet of *H. americanus* was substantially higher. In the laboratory, *H. americanus* exhibited selective feeding behavior.

In arthropods, the growth of new tissue occurs continuously but only becomes visible when the animal molts, shedding its old shell during the process known as ecdysis. The majority of the growth in length and weight happens during ecdysis. The entire external body shell, including the largest parts like the carapace and legs and the smallest parts like the tactile and chemosensory hairs, as well as the chitinous lining of the esophagus and foregut and the extremely thin surface of the gills, are all included in the exoskeleton that has been abandoned. A spectacular metamorphosis occurs during molting, which occurs at the conclusion of the larval stage. It is the transition from the leaf-shaped phyllosoma larva to the adult-shaped puerulus in spiny lobsters. The fourth larval molt in nephropid lobsters is accompanied by a transition from the fourth stage's adult body form to one that resembles a shrimp. A lobster's entire existence is devoted to either getting ready for the subsequent molt or recovering from the previous one. There are various stages in the period between molts that can be distinguished morphologically and physiologically. There are essentially two hormones or hormone classes. One is known as the molt-inhibiting hormone (MIH) and is produced and stored in the X-organ sinus gland complex of the eyestalk. The molting hormone (MH), the other, is most likely created in a molting gland. Which tissue in the lobster cephalothorax serves as the molting gland is not yet obvious from the evidence.

According to current theories, ß-ecdysone mediates changes in the epidermis and cuticle during the later phases of premolt, whereas ecdysone may promote premolt. Due to the rigidity of the exoskeleton, growth, which is a nearly continual process, can only be manifested in size increase during the time of molting. The length of time between each ecdysis and the subsequent one and the size increase at ecdysis determine a lobster's growth rate. Size growth is often quantified by body weight or carapace length. In its first year, *Homarus americanus* may molt up to ten times; in its second and third years, it may molt three or four times; in its third and fourth years, it may molt twice; and in the years following, it molts just once a year. When it reaches a large size, Homarus will molt fewer than once every year. Similar slowing may be seen in the length relationship, where aging results in a reduction in the percentage increase in carapace length during molt. As a result, as lobsters age, they do not molt as quickly or expand in size by the same percentage at each molt. The most significant environmental component influencing growth rate is temperature. The growth rate of *Panulirus cygnus* kept in heated water year-round has also increased. A shorter intermolt period, with the percentage of size increase remaining the same or slightly declining, accounts for the enhanced growth rate. Nutritional status affects growth rate as well, most likely through lengthening the interval between molts. Intermolt time is also influenced by the presence or absence of additional individuals of the same species. In contrast to the normally solitary *H. americanus*, *P. cygnus* grows more slowly when raised individually compared to when raised in groups. In other words, animals raised in groups develop more slowly than those raised alone. The growth rate is also influenced by other elements, including oxygen concentration, container size and time of year. Additionally, the growth rate is affected by intrinsic variables. Both spiny and clawed lobsters experience slower growth rates as a result of losing some of their appendages and having them grow back later.

Most lobsters have well-developed autotomy, or the capacity to remove an appendage by breaking it at a predefined specialized place. It appears to be a reaction that enables escape from a predator's clutches and happens when a claw or leg is intensely aroused. When *Panulirus cygnus* is captured or removed from a tank, it has been reported to autotomize one to several legs. Contrarily, Homarus rarely autotomizes the big chelae unless highly stimulated and the remaining pereiopods are significantly more difficult to induce autotomy in. If molting is sluggish and a leg becomes caught in the old exoskeleton, autotomy may also happen. The likelihood of having an appendage amputated diminishes with age and the number of lost limbs. Additionally, the likelihood of autotomy is influenced by the animal's health and how long it has been in captivity. Two levator muscles, which are also used for the animal's typical gait, contract in an autonomic response to gross sensory input. When the limb is vigorously stimulated, the smaller of the two muscles will contract, rupturing a tendon of the larger. In place of the bigger muscle, a small skeletal plug that crosses the breaking plane now receives the larger muscle's application. A slight force applied externally to the limb is sufficient to remove the plug once the bigger muscle pulls it out of its socket distal to the breaking plane.

The body will restore various components after injury or autotomy. A limb bud begins to form in the case of a lost appendage soon after the loss, but proecdysis is when it grows the fastest. Particularly while claws are being regenerated, the entire appendage appears fully developed but smaller than usual at the following molt. Depending on what stage of the molt cycle the damage occurs at, the regenerating appendage will vary in size and appearance. The limb grows to its full size after a number of molts. In *Panulirus cygnus*, *P. interruptus*, or *Homarus americanus*, the autotomy of numerous legs will shorten the intermolt interval and lessen the size increment at the following molt. The mouthparts of many decapods do not regenerate and lobsters appear to be one of these animals. If an eyestalk is removed, it won't grow again, but every so often an antenna or leg-like heteromorphic appendage will take its place. The ovaries are paired cylindrical rods dorsal and lateral to the digestive tract, linked by a transverse bridge under the heart and follow the overall decapod pattern. They extend to the insertion of the abdomen or beyond from the anterior end of the stomach, roughly at eye level. The ovary and the vaginal opening on the third leg's coxa are connected by a little channel. Creamy white in appearance, the juvenile ovary eventually turns green in Homarus and *Nephrops norvegicus*, orange and then red in Jasus and Panulirus and royal blue in *Metanephrops andamanicus*.

After reaching maturity, Homarus only spawns once every two years, but Nephrops spawns annually. The management of the reproductive cycle is poorly understood. The molting hormones MIH and MH are thought to be among the hormones that are at play. Temperature, seasonal variables and perhaps photoperiod are also involved. In Homarus, the ovary is reabsorbed and the molt cycle advances if the molt cycle and the ovarian development cycle are out of sync, or if the eggs are extruded too late in proecdysis, causing the female to molt before the eggs hatch. The blood is a bright green tint due to pigments from the reabsorbing ovaries. When lobsters are kept in particularly unfavorable conditions, reabsorption occasionally happens later in the cycle, causing the hemolymph to turn a very dark green and creating a so-called black lobster. The primary causes of mortality in any wild population are disease and parasites, although little is known about the diseases that affect lobsters outside of what has been discovered via attempts to keep or cultivate the animals. When the water quality is good, lobsters appear to be incredibly resilient and resistant to many different sorts of diseases.

The term shell disease refers to a pitted appearance of the shell brought on by chitinoclastic microorganisms destroying the exoskeleton. At least in *H. americanus*, the disease seems to have a bacterial etiology. Except in highly contaminated locations, the disease is uncommon in natural populations. However, it can become severe in lobsters kept in impoundments for extended periods of time. No alternative to maintaining a decent environmental quality has been proposed as a treatment. When the gill membranes are fouled by the common filamentous bacterium *Leucothrix mucor*, anoxia results or the exoskeleton is not completely shed during molting, which results in significant mortality in the larval stages. Leucothrix and other filamentous bacteria obtain their nutrition from the water around them and their growth is accelerated as the amount of dissolved organic matter in the water increases. For the treatment of infections, penicillin and streptomycin have been suggested, however, this permits other germs to invade, frequently with equally damaging outcomes. There have been several antifungal drugs proposed for management, but the best course of action is probably to carefully clean the culture system and maintain low nutrient contents.

There is one more incapacitating condition that should be mentioned here, even though it is not a disease. The circulating water system occasionally contains tiny gas bubbles that are small enough to penetrate the integument of the gill surfaces and enter the bloodstream. The sole therapy is to eliminate the cause of the air bubbles, which may be a leak on the pumping system's suction side or the outcome of an excessively quick increase in water temperature. Another method for eliminating air bubbles is to cascade water over a number of baffles. The clawed lobsters typically choose solitary shelter and protect it from other species and members of their own. As opposed to vertebrates, Homarus does not exhibit territorial boundary behaviors in its aggressive repertoire. However, there are periods when a region around the burrow is protected, such as when the molt is approaching and during courtship. Both adult and juvenile spiny lobsters share shelters, though to a lower extent. When spiny lobsters are in their dens, aggressive behavior is reduced, allowing vast numbers of species to congregate. Although lobsters' reproductive behavior varies quite a little in specifics, the overall pattern is consistent. Courtship involves chemical, physical and visual communication and can last anywhere from a few hours to several days. The sperm packet is placed either inside or on the ventral surface of the female during belly-to-belly copulation. Days or months later, the eggs are fertilized, extruded and cemented to the swimmerets. All Nephropidae, as well as some but not all spiny lobsters, mate when the female has just undergone molting. Many biological, economic and bioeconomic models created for lobster fisheries management have been created to help address these issues. Unfortunately, most of these models were not created with the mixing of the two disciplines in mind; rather, they were created from a solely economic or biological perspective. For instance, compared to that of insects, the metabolism of crustaceans is fairly poorly understood; other, less extreme examples include endocrinology and nutrition. But there is no requirement for such a circumstance. Many of the bigger decapod Crustacea species are easily accessible, resilient and maintainable in captivity, which makes them ideal experimental animals. These Crustacea are more manageable in a lot of ways than insects because of their greater size.

References

1. Acheson, J. (2001). Confounding the Goals of Management: Response of the Maine Lobster Industry to a Trap Limit. *North American Journal of Fisheries Management*, *21*, 404–416. https://doi.org/10.1577/1548-8675(2001)021<0404:CTGOMR>2.0.CO;2
2. Alborés, I., García-Soler, C., & Fernández, L. (2019). Reproductive biology of the slipper lobster Scyllarus arctus in Galicia (NW Spain): Implications for fisheries management. *Fisheries Research*, *212*, 1–11. https://doi.org/10.1016/j.fishres.2018.12.001
3. Ambati, R. R., Siew Moi, P., Ravi, S., & Aswathanarayana, R. G. (2014). Astaxanthin: Sources, Extraction, Stability, Biological Activities and Its Commercial Applications—A Review. *Marine Drugs*, *12*(1), 128–152. https://doi.org/10.3390/md12010128
4. Barnett, A., & Anderies, J. (2014). Weak feedbacks, governance mismatches and the robustness of social-ecological systems: An analysis of the Southwest Nova Scotia lobster fishery with comparison to Maine. *Ecology and Society*, *19*(4). https://doi.org/10.5751/ES-06714-190439
5. Barth, H.-J., & Khan, N. (2008). Biogeophysical setting of the Gulf. In *Protecting the Gulf's Marine Ecosystems from Pollution* (pp. 1–21). https://doi.org/10.1007/978-3-7643-7947-6_1
6. Bermudes, M., & Ritar, A. (2008). Response of early stage spiny lobster *Jasus edwardsii* phyllosoma larvae to changes in temperature and photoperiod. *Aquaculture*, *281*, 63–69. https://doi.org/10.1016/j.aquaculture.2008.05.035

7. Bowser, P. R., Rosemark, R., & Reiner, C. R. (1981). A preliminary report of vibriosis in cultured American lobsters, *Homarus americanus*. *Journal of Invertebrate Pathology*, *37*(1), 80–85. https://doi.org/10.1016/0022-2011(81)90058-6
8. Breen, P. (2009). A voluntary harvest control rule for a New Zealand rock lobster (*Jasus edwardsii*) stock. *New Zealand Journal of Marine and Freshwater Research*, *43*, 941–951. https://doi.org/10.1080/00288330909510052
9. Burrowing—An overview | ScienceDirect Topics. (n.d.). Retrieved June 29, 2022, from https://www.sciencedirect.com/topics/agricultural-and-biological-sciences/burrowing
10. Butler, M. J., IV. (2017). Collecting and processing lobsters. *Journal of Crustacean Biology*, *37*(3), 340–346. https://doi.org/10.1093/jcbiol/rux021
11. Butler, M. J., Steneck, R. S., & Herrnkind, W. F. (2006). Juvenile and Adult Ecology. In *Lobsters: Biology, Management, Aquaculture and Fisheries* (pp. 263–309). John Wiley & Sons, Ltd. https://doi.org/10.1002/9780470995969.ch8
12. Canli, M., & Stagg, R. M. (1996). The effects of in vivo exposure to cadmium, copper and zinc on the activities of gill ATPases in the Norway lobster, Nephrops norvegicus. *Archives of Environmental Contamination and Toxicology*, *31*(4), 494–501. https://doi.org/10.1007/BF00212433
13. Carlberg, J. M., Van Olst, J. C., & Ford, R. F. (1978). A Comparison Of Larval And Juvenile Stages Of The Lobsters *Homarus americanus*, *Homarus gammarus* and Their Hybrid1. *Proceedings of the Annual Meeting - World Mariculture Society*, *9*(1–4), 109–122. https://doi.org/10.1111/j.1749-7345.1978.tb00236.x
14. Cetina-Heredia, P., Roughan, M., Sebille, E., Feng, M., & Coleman, M. (2015). Strengthened currents override the effect of warming on lobster larval dispersal & survival. *Global Change Biology*, *21*. https://doi.org/10.1111/gcb.13063
15. Chang, S.-C., Chan, T.-Y., & Kumar, A. B. (2020). Deep-sea clawed lobster *Nephropsis stewarti* Wood-Mason, 1872 species complex in the Indo-West Pacific (Crustacea, Decapoda, Nephropidae), with description of a new species. *ZooKeys*, *1008*, 37–60. https://doi.org/10.3897/zookeys.1008.59966

16. Cheng, J. H., & Chang, E. S. (1991). Ecdysteroid Treatment Delays Ecdysis in the Lobster, *Homarus americanus*. *The Biological Bulletin*, *181*(1), 169–174. https://doi.org/10.2307/1542499
17. Chubb, C. f. (2000). Reproductive Biology: Issues for Management. In *Spiny Lobsters* (pp. 243–275). John Wiley & Sons, Ltd. https://doi.org/10.1002/9780470698808.ch14
18. Coutures, E. (2000). Distribution of phyllosoma larvae of Scyllaridae and Palinuridae (Decapoda: Palinuridea) in the south-western lagoon of New Caledonia. *Marine and Freshwater Research*, *51*, 363–369. https://doi.org/10.1071/MF98108
19. Crear, B. J., Thomas, C. W., Hart, P. R., & Carter, C. G. (2000). Growth of juvenile southern rock lobsters, *Jasus edwardsii*, is influenced by diet and temperature, whilst survival is influenced by diet and tank environment. *Aquaculture*, *190*(1), 169–182. https://doi.org/10.1016/S0044-8486(00)00391-4
20. Cruz, R., T. Torres, M., Santana, J. V. M., & Cintra, I. H. A. (2021). Lobster Distribution and Biodiversity on the Continental Shelf of Brazil: A Review. *Diversity*, *13*(11), 507. https://doi.org/10.3390/d13110507
21. De Grave, S., Pentcheff, N. D., Ahyong, S. T., Chan, T.-Y., Crandall, K. A., Dworschak, P. C., … Wetzer, R. (2009). *A Classification of Living and Fossil Genera of Decapod Crustaceans*. Retrieved from http://repository.si.edu/xmlui/handle/10088/8358
22. Derby, C., & Atema, J. (2011). Influence of Drilling Muds on the Primary Chemosensory Neurons in Walking Legs of the Lobster, *Homarus americanus*. *Canadian Journal of Fisheries and Aquatic Sciences*, *38*, 268–274. https://doi.org/10.1139/f81-038
23. DeVantier, L., Turak, E., & Szava-Kovats, R. (2020). Species Richness and Abundance of Reef-Building Corals in the Indo-West Pacific: The Local–Regional Relation Revisited. *Frontiers in Marine Science*, *7*. Retrieved from https://www.frontiersin.org/article/10.3389/fmars.2020.00487
24. Dow, R. L. (1980). Chapter 8—The Clawed Lobster Fisheries. In J. S. Cobb & B. F. Phillips (Eds.), *The Biology and Management of Lobsters* (pp. 265–316). San Diego: Academic Press. https://doi.org/10.1016/B978-0-08-091734-4.50016-0
25. Ehrhardt, N. (2005). Population Dynamic Characteristics and

Sustainability Mechanisms in Key Western Central Atlantic Spiny Lobster, Panulirus Argus, Fisheries. *Bulletin of Marine Science*, *76*, 501–526.

26. Ehrhardt, N. M. (2008). Estimating growth of the Florida spiny lobster, *Panulirus argus*, from molt frequency and size increment data derived from tag and recapture experiments. *Fisheries Research*, *93*(3), 332–337. https://doi.org/10.1016/j.fishres.2008.06.008
27. Elner, R. W., & Campbell, A. (1981). Force, function and mechanical advantage in the chelae of the American lobster *Homarus americanus* (Decapoda: Crustacea). *Journal of Zoology*, *193*(2), 269–286. https://doi.org/10.1111/j.1469-7998.1981.tb03444.x
28. Faria-Junior, E., Gaeta, J., & Freire, A. (2013). An update on the lobster species (Panulirus White, 1847) from the Abrolhos Marine National Park, Northern Brazil. *Revista de Investigaciones Marinas*, *33*.
29. Farmaki, E., Thomaidis, N., Pasias, I., Baulard, C., Papaharisis, L., & Efstathiou, C. (2014). Environmental impact of intensive aquaculture: Investigation on the accumulation of metals and nutrients in marine sediments of Greece. *The Science of the Total Environment*, *485–486*, 554–562. https://doi.org/10.1016/j.scitotenv.2014.03.125
30. Ferran, P., Rebeca, G., Michael, H., & Paul, C. (2016). DNA barcoding the phyllosoma of *Scyllarides squammosus* (H. Milne Edwards, 1837) (Decapoda: Achelata: Scyllaridae). *Frontiers in Marine Science*, *3*. https://doi.org/10.3389/conf.FMARS.2016.05.00131
31. Fitzgibbon, Q. P., & Battaglene, S. C. (2012). Effect of water temperature on the development and energetics of early, mid and late-stage phyllosoma larvae of spiny lobster *Sagmariasus verreauxi*. *Aquaculture*, *344–349*, 153–160. https://doi.org/10.1016/j.aquaculture.2012.03.008
32. Fournier, D. A., Sibert, J. R., & Terceiro, M. (1991). Analysis of Length Frequency Samples with Relative Abundance Data for the Gulf of Maine Northern Shrimp (*Pandalus borealis*) by the MULTIFAN Method. *Canadian Journal of Fisheries and Aquatic Sciences*, *48*(4), 591–598. https://doi.org/10.1139/f91-075
33. Frank, T., & Case, J. (1988). Visual Spectral Sensitivities of Bioluminescent Deep-Sea Crustaceans. *Biological Bulletin*, *175*, 261. https://doi.org/10.2307/1541567
34. Greenan, B. J. W., Shackell, N. L., Ferguson, K., Greyson, P.,

Cogswell, A., Brickman, D., … Saba, V. S. (2019). Climate Change Vulnerability of American Lobster Fishing Communities in Atlantic Canada. *Frontiers in Marine Science*, *6*. Retrieved from https://www.frontiersin.org/article/10.3389/fmars.2019.00579

35. Groeneveld, J. C., Greengrass, C. L., Zyl, D. van, & Branch, G. M. (2010). Settlement patterns, size and growth of puerulus and juvenile rock lobster *Jasus lalandii* at an oyster farm in Saldanha Bay, South Africa. *African Journal of Marine Science*, *32*(3). Retrieved from https://www.ajol.info/index.php/ajms/article/view/62774
36. Groeneveld, Johan C., Gopal, K., George, R. W., & Matthee, C. A. (2007). Molecular phylogeny of the spiny lobster genus Palinurus (Decapoda: Palinuridae) with hypotheses on speciation in the NE Atlantic/Mediterranean and SW Indian Ocean. *Molecular Phylogenetics and Evolution*, *45*(1), 102–110. https://doi.org/10.1016/j.ympev.2007.06.002
37. Harrington, A., Tudor, M., Reese, H., Bouchard, D., & Hamlin, H. (2019). Effects of temperature on larval American lobster (*Homarus americanus*): Is there a trade-off between growth rate and developmental stability? *Ecological Indicators*, *96*, 404–411. https://doi.org/10.1016/j.ecolind.2018.09.022
38. Haryono, F. E. D., Ambariyanto, & Sulistyo, I. (2018). Sex Diversity Approach of Spiny Lobster ($\less$i$\greater$Panulirus$\less$/i$\greater$spp) to Marine Oil Spill Pollution in Southern Waters of Java. *IOP Conference Series: Earth and Environmental Science*, *116*, 012008. https://doi.org/10.1088/1755-1315/116/1/012008
39. Hild, S., Marti, O., & Ziegler, A. (2008). Spatial distribution of calcite and amorphous calcium carbonate in the cuticle of the terrestrial crustaceans *Porcellio scaber* and *Armadillidium vulgare*. *Journal of Structural Biology*, *163*(1), 100–108. https://doi.org/10.1016/j.jsb.2008.04.010
40. Howell, P. (2012). The Status of the Southern New England Lobster Stock. *Journal of Shellfish Research*, *31*, 573–579. https://doi.org/10.2983/035.031.0217
41. Iv, M. J. B., Jr, D. C. B., Iii, T. W. D., Moss, J., & Shields, J. D. (2015). Behavioral Immunity Suppresses an Epizootic in Caribbean Spiny Lobsters. *PLOS ONE*, *10*(6), e0126374. https://doi.org/10.1371/journal.pone.0126374

42. Jeena, N. S., Gopalakrishnan, A., Radhakrishnan, E. V., Kizhakudan, J. K., Basheer, V. S., Asokan, P. K., & Jena, J. K. (2016). Molecular phylogeny of commercially important lobster species from Indian coast inferred from mitochondrial and nuclear DNA sequences. *Mitochondrial DNA Part A*, *27*(4), 2700–2709. https://doi.org/10.3109/19401736.2015.1046160
43. Jeffs, A. G., Nichols, P. D., & Bruce, M. P. (2001). Lipid reserves used by pueruli of the spiny lobster *Jasus edwardsii* in crossing the continental shelf of New Zealand. *Comparative Biochemistry and Physiology Part A: Molecular & Integrative Physiology*, *129*(2), 305–311. https://doi.org/10.1016/S1095-6433(00)00348-2
44. Jeffs, A., & James, P. (2001). Sea-cage culture of the spiny lobster *Jasus edwardsii* in New Zealand. *Marine and Freshwater Research*, *52*, 1419–1424. https://doi.org/10.1071/MF01064
45. Jones, C. (2007). *Improving lobster grow-out and nutrition in Nusa Tenggara Barat – a feasibility study. Final Report for project SMAR/2007/228*.
46. Jones, C., Linton, L., Horton, D., & Bowman, W. (2001). Effect of density on growth and survival of ornate rock lobster, *Panulirus ornatus* (Fabricius, 1798), in a flow-through raceway system. *Marine and Freshwater Research*, *52*, 1425–1429. https://doi.org/10.1071/MF01053
47. Jones, C., & Shanks, S. (2009). *Requirements for the aquaculture of Panulirus ornatus in Australia*.
48. Kampouris, T. E., Asimaki, A., Klaoudatos, D., Exadactylos, A., Karapanagiotidis, I. T., & Batjakas, I. E. (2021). Nutritional Quality of the European Spiny Lobster *Palinurus elephas* (J.C. Fabricius, 1787) (Achelata, Palinuridae) and the Non-Indigenous Northern Brown Shrimp Penaeus aztecus Ives, 1891 (Dendrobranchiata, Penaeidae). *Foods*, *10*(10), 2480. https://doi.org/10.3390/foods10102480
49. Kavanaugh, M. T., Rheuban, J. E., Luis, K. M. A., & Doney, S. C. (2017). Thirty-Three Years of Ocean Benthic Warming Along the U.S. Northeast Continental Shelf and Slope: Patterns, Drivers and Ecological Consequences. *Journal of Geophysical Research. Oceans*, *122*(12), 9399–9414. https://doi.org/10.1002/2017JC012953
50. Kittaka, J., Kudo, R., Onoda, S., Kanemaru, K., & Mercer, J. (2001). Larval culture of the European spiny lobster *Palinurus elephas*. *Marine*

and Freshwater Research, *52*, 1439–1444. https://doi.org/10.1071/MF01188

51. Kommuri, P. K., Mugada, N., & Kondamudi, R. B. (2021). Proximate Analysis and Mineral Composition of Commercially Important Spiny Lobsters from Visakhapatnam Coast, Andhra Pradesh, India. *Asian Journal of Fisheries and Aquatic Research*, 39–47. https://doi.org/10.9734/ajfar/2021/v14i230294
52. Lavalli, K., & Spanier, E. (2007). *Introduction to the Biology and Fisheries of Slipper Lobsters*. https://doi.org/10.1201/9781420005165.pt1
53. Lavalli, K., Spanier, E., & Goldstein, J. (2019). *Scyllarid Lobster Biology and Ecology*. https://doi.org/10.5772/intechopen.88218
54. Lavalli, K., Spanier, E., & Grasso, F. (2007). Behavior and Sensory Biology of Slipper Lobsters. In *The Biology and Fisheries of the Slipper Lobster* (pp. 133–181). https://doi.org/10.1201/9781420005165.ch7
55. Lee, S., Hartstein, N., Yee, R. W., & Jeffs, A. (2014). Assessment of the production and dispersal of faecal waste from the sea-cage aquaculture of spiny lobsters. *Aquaculture Research*, *47*. https://doi.org/10.1111/are.12618
56. Lenihan, H. S., Gallagher, J. P., Peters, J. R., Stier, A. C., Hofmeister, J. K. K., & Reed, D. C. (2021). Evidence that spillover from Marine Protected Areas benefits the spiny lobster (*Panulirus interruptus*) fishery in southern California. *Scientific Reports*, *11*, 2663. https://doi.org/10.1038/s41598-021-82371-5
57. L.H., E., & Jones, J. (1999). *International Symposium on lobster Health Management: Proceedings*.
58. Li, C., Shields, J. D., Small, H. J., Reece, K. S., Hartwig, C. L., Cooper, R. A., & Ratzlaff, R. E. (2006). Detection of Panulirus argus Virus 1 (PaVI) in the Caribbean spiny lobster using fluorescence in situ hybridization (FISH). *Diseases of Aquatic Organisms*, *72*(3), 185–192. https://doi.org/10.3354/dao072185
59. Lin, W., Ren, Z., Mu, C., Ye, Y., & Wang, C. (2020). Effects of Elevated pCO2 on the Survival and Growth of *Portunus trituberculatus*. *Frontiers in Physiology*, *11*, 750. https://doi.org/10.3389/fphys.2020.00750
60. Luypaert, T., Hagan, J. G., McCarthy, M. L., & Poti, M. (2020). Status of Marine Biodiversity in the Anthropocene. In S. Jungblut, V. Liebich, & M. Bode-Dalby (Eds.), *YOUMARES 9—The Oceans: Our Research,*

Our Future: Proceedings of the 2018 conference for YOUng MArine RESearcher in Oldenburg, Germany (pp. 57–82). Cham: Springer International Publishing. https://doi.org/10.1007/978-3-030-20389-4_4

61. Ma, K. Y., Chan, T.-Y., & Chu, K. H. (2011). Refuting the six-genus classification of *Penaeus* s.l. (Dendrobranchiata, Penaeidae): A combined analysis of mitochondrial and nuclear genes. *Zoologica Scripta*, *40*(5), 498–508. https://doi.org/10.1111/j.1463-6409.2011.00483.x
62. Mahdi Abkener, A. (2020). Assessment of heavy metals pollution in muscle of sole (*Cynoglossus arel*), spiny lobster (*Panulirus homarus*) and sediments in the northern coasts of the Oman Sea during pre and post monsoon. *Iranian Journal of Fisheries Sciences*, *19*(4), 1638–1656. https://doi.org/10.22092/ijfs.2018.124017.1006
63. Manambrakat, V., Venkatesan, R., Senthil, T., Kumar, M., Jha, D., Remany, M., … Kathiroli, S. (2009). Farming of spiny lobsters in sea cages in India Farming of spiny lobsters in sea cages in India. *New Zealand Journal of Marine and Freshwater Research*, *43*, 623–634.
64. Marchese, G., Fitzgibbon, Q. P., Trotter, A. J., Carter, C. G., Jones, C. M., & Smith, G. G. (2019). The influence of flesh ingredients format and krill meal on growth and feeding behaviour of juvenile tropical spiny lobster *Panulirus ornatus*. *Aquaculture*, *499*, 128–139. https://doi.org/10.1016/j.aquaculture.2018.09.019
65. Matzen da Silva, J., Creer, S., dos Santos, A., Costa, A. C., Cunha, M. R., Costa, F. O., & Carvalho, G. R. (2011). Systematic and Evolutionary Insights Derived from mtDNA COI Barcode Diversity in the Decapoda (Crustacea: Malacostraca). *PLoS ONE*, *6*(5), e19449. https://doi.org/10.1371/journal.pone.0019449
66. McLeay, L. J., Doubell, M. J., & Linnane, A. J. (2019). Spatial and temporal variations in female size at maturity of a Southern Rock Lobster (Jasus edwardsii) population: A likely response to climate change. *PLoS ONE*, *14*(11), e0225144. https://doi.org/10.1371/journal.pone.0225144
67. Medina, A., Vila, Y., Megina, C., Sobrino, I., & Ramos, F. (2000). A Histological Study Of The Age-pigment, Lipofuscin, In Dendrobranchiate Shrimp Brains. *Journal of Crustacean Biology*, *20*(3), 423–430. https://doi.org/10.1651/0278-0372(2000)020[0423:AHSOTA]2.0.CO;2

68. Meyer, D., Fonseca, M., Murphey, P., Mcmichael, R., Michael, J., Byerly, M. M., … Thayer, G. (1998). Effects of live-bait shrimp trawling on seagrass beds and fish bycatch in Tampa Bay, Florida. *Fish. Bull*, *97*, 193–199.

69. Militz, T. A., Leini, E., Duy, N. D. Q., & Southgate, P. C. (2018). Successful large-scale hatchery culture of sandfish (*Holothuria scabra*) using micro-algae concentrates as a larval food source. *Aquaculture Reports*, *9*, 25–30. https://doi.org/10.1016/j.aqrep.2017.11.005

70. Miller, M. R., Pearce, L., & Bettjeman, B. I. (2014). Detailed Distribution of Lipids in Greenshell™ Mussel (*Perna canaliculus*). *Nutrients*, *6*(4), 1454–1474. https://doi.org/10.3390/nu6041454

71. Monnereau, I., & Pollnac, R. (2012). Which Fishers are Satisfied in the Caribbean? A Comparative Analysis of Job Satisfaction Among Caribbean Lobster Fishers. *Social Indicators Research*, *109*, 95–118. https://doi.org/10.1007/s11205-012-0058-0

72. Moss, G., James, P., Allen, S., & Bruce, M. (2004). Temperature effects on the embryo development and hatching of the spiny lobster *Sagmariasus verreauxi*. *New Zealand Journal of Marine and Freshwater Research - N Z J MAR FRESHWATER RES*, *38*, 795–801. https://doi.org/10.1080/00288330.2004.9517278

73. Nakatsuji, T., Lee, C.-Y., & Watson, R. D. (2009). Crustacean molt-inhibiting hormone: Structure, function and cellular mode of action. *Comparative Biochemistry and Physiology Part A: Molecular & Integrative Physiology*, *152*(2), 139–148. https://doi.org/10.1016/j.cbpa.2008.10.012

74. Nankervis, L., & Jones, C. (n.d.). Recent advances and future directions in practical diet formulation and adoption in tropical Palinurid lobster aquaculture. *Reviews in Aquaculture*, *n/a*(n/a). https://doi.org/10.1111/raq.12675

75. Nguyen, L. (2009). *Economic analysis of the environmental impact on marine cage lobster aquaculture in Vietnam*.

76. Nigam, P. K., & Nigam, A. (2010). Botulinum Toxin. *Indian Journal of Dermatology*, *55*(1), 8–14. https://doi.org/10.4103/0019-5154.60343

77. Norman-López, A., Pascoe, S., Thébaud, O., van putten, I., Innes, J., Jennings, S., … Plaganyi, E. (2014). Price integration in the Australian rock lobster industry: Implications for management and climate change adaptation. *Australian Journal of Agricultural and Resource Economics*, *In Press*. https://doi.org/10.1111/1467-8489.12020

78. Paulus, E. (2021). Shedding Light on Deep-Sea Biodiversity—A Highly Vulnerable Habitat in the Face of Anthropogenic Change. *Frontiers in Marine Science*, *8*. Retrieved from https://www.frontiersin.org/article/10.3389/fmars.2021.667048

79. Pere, A., Marengo, M., Lejeune, P., & Durieux, E. (2019). Evaluation of Homarus gammarus (Crustacea: Decapoda: Nephropidae) catches and potential in a Mediterranean small-scale fishery. *Scientia Marina*, *83*, 0–000. https://doi.org/10.3989/scimar.04862.22B

80. Perez Benavente, G., Uglem, I., Browne, R., & Marino Balsa, C. (2010). Culture of juvenile European lobster (*Homarus gammarus* L.) in submerged cages. *Aquaculture International*, *18*(6), 1177–1189. https://doi.org/10.1007/s10499-010-9332-9

81. Plaganyi, E., Mcgarvey, R., Gardner, C., Caputi, N., Dennis, D., De Lestang, S., … Villanueva, C. (2018). Overview, opportunities and outlook for Australian spiny lobster fisheries. *Reviews in Fish Biology and Fisheries*, *28*. https://doi.org/10.1007/s11160-017-9493-y

82. Priyambodo, B., Jones, C., & Sammut, J. (2020). Assessment of the lobster puerulus (*Panulirus homarus* and *Panulirus ornatus*, Decapoda: Palinuridae) resource of Indonesia and its potential for sustainable harvest for aquaculture. *Aquaculture*, *528*, 735563. https://doi.org/10.1016/j.aquaculture.2020.735563

83. Qadri, S. A., Camacho, J., Wang, H., Taylor, J. R., Grosell, M., & Worden, M. K. (2007). Temperature and acid-base balance in the American lobster *Homarus americanus*. *The Journal of Experimental Biology*, *210*(Pt 7), 1245–1254. https://doi.org/10.1242/jeb.02709

84. Radhakrishnan, E. V., Kizhakudan, J. K., & Phillips, B. F. (2019). Introduction to Lobsters: Biology, Fisheries and Aquaculture. In E. V. Radhakrishnan, B. F. Phillips, & G. Achamveetil (Eds.), *Lobsters: Biology, Fisheries and Aquaculture* (pp. 1–33). Singapore: Springer. https://doi.org/10.1007/978-981-32-9094-5_1

85. Rafael, P., Vázquez, S., Martinez, J., & de León, M. (2005). Bioeconomic modelling and risk assessment of the Cuban fishery for spiny lobster *Panulirus argus*. *Fisheries Research*, *75*, 149–163. https://doi.org/10.1016/j.fishres.2005.03.014

86. Rajendiran, S., & Vasudevan, S. (2016). Localization and identification of crustacean hyperglycemic hormone producing neurosecretory cells in the eyestalk of blue swimmer crab, *Portunus pelagicus*. *Microscopy*

Research and Technique, *79*(11), 1024–1030. https://doi.org/10.1002/jemt.22737

87. Ritar, A. (2001). The experimental culture of phyllosoma larvae of southern rock lobster (*Jasus edwardsii*) in a flow-through system. *Aquacultural Engineering*, *24*, 149–156. https://doi.org/10.1016/S0144-8609(00)00072-8
88. Ritchie, S., Cortis, G., Paton, C., Townsend, M., Shroyer, D., Zborowski, P., … van den Hurk, A. (2013). A Simple Non-Powered Passive Trap for the Collection of Mosquitoes for Arbovirus Surveillance. *Journal of Medical Entomology*, *50*, 185–194. https://doi.org/10.1603/ME12112
89. Rochette, R., Sainte-Marie, B., Allain, M., Baker, J., Bernatchez, L., Boudreau, V., … Tremblay, M. J. (2018). The Lobster Node of the CFRN: Co-constructed and collaborative research on productivity, stock structure and connectivity in the American lobster (*Homarus americanus*). *Canadian Journal of Fisheries and Aquatic Sciences*, *75*(5), 813–824. https://doi.org/10.1139/cjfas-2016-0426
90. Salisbury, J. E., & Jönsson, B. F. (2018). Rapid warming and salinity changes in the Gulf of Maine alter surface ocean carbonate parameters and hide ocean acidification. *Biogeochemistry*, *141*(3), 401–418. https://doi.org/10.1007/s10533-018-0505-3
91. Sánchez-Paz, A. (2010). White spot syndrome virus: An overview on an emergent concern. *Veterinary Research*, *41*(6), 43. https://doi.org/10.1051/vetres/2010015
92. Shields, J., Wheeler, K., & Moss, J. (2012). Histological Assessment of the Lobster (*Homarus americanus*) in the "100 Lobsters" Project. *Journal of Shellfish Research*, *31*, 439–447. https://doi.org/10.2983/035.031.0204
93. Singh, S. P., Groeneveld, J. C., Hart-Davis, M. G., Backeberg, B. C., & Willows-Munro, S. (2018). Seascape genetics of the spiny lobster *Panulirus homarus* in the Western Indian Ocean: Understanding how oceanographic features shape the genetic structure of species with high larval dispersal potential. *Ecology and Evolution*, *8*(23), 12221–12237. https://doi.org/10.1002/ece3.4684
94. Spanier, E., Lavalli, K. L., & Edelist, D. (2010). Artificial habitats for benthic dwelling lobsters—Analysis of 5 decades of research. *Journal of the Marine Biological Association of India*, 26.
95. Steell, C., Cooke, S., & Eliason, E. (2020). Artificial light at night does

not alter heart rate or locomotor behaviour in Caribbean spiny lobster (*Panulirus argus*): Insights into light pollution and physiological disturbance using biologgers. *Conservation Physiology*, *8*. https://doi.org/10.1093/conphys/coaa097

96. Stein, L., Jacobson, S., & Atema, J. (1975). *Behavior of lobsters (Homarus americanus) in a semi-natural environment at ambient temperatures and under thermal stress* [Technical Report]. Woods Hole Oceanographic Institution. https://doi.org/10.1575/1912/3945
97. Stewart, J., Cornick, J., Spears, D., & McLeese, D. (2011). Incidence of *Gaffkya homari* in Natural Lobster (*Homarus americanus*) Populations of the Atlantic Region of Canada. *Journal of the Fisheries Research Board of Canada*, *23*, 1325–1330. https://doi.org/10.1139/f66-123
98. Stewart, J. E. (1984). Lobster diseases. *Helgoländer Meeresuntersuchungen*, *37*(1), 243–254. https://doi.org/10.1007/BF01989309
99. Stoll, J. S., Crona, B. I., Fabinyi, M., & Farr, E. R. (2018). Seafood Trade Routes for Lobster Obscure Teleconnected Vulnerabilities. *Frontiers in Marine Science*, *5*. Retrieved from https://www.frontiersin.org/article/10.3389/fmars.2018.00239
100. Taylor, C. M., Golding, J., & Emond, A. M. (2016). Blood mercury levels and fish consumption in pregnancy: Risks and benefits for birth outcomes in a prospective observational birth cohort. *International Journal of Hygiene and Environmental Health*, *219*(6), 513–520. https://doi.org/10.1016/j.ijheh.2016.05.004
101. Thomas, C. W., Crear, B. J., & Hart, P. R. (2000). The effect of temperature on survival, growth, feeding and metabolic activity of the southern rock lobster, *Jasus edwardsii*. *Aquaculture*, *185*(1), 73–84. https://doi.org/10.1016/S0044-8486(99)00341-5
102. Tlusty, M. F., Myers, A., & Metzler, A. (2008). Short- and long-term dietary effects on disease and mortality in American lobster *Homarus americanus*. *Diseases of Aquatic Organisms*, *78*(3), 249–253. https://doi.org/10.3354/dao01867
103. Torres, A. P., Palero, F., Dos Santos, A., Abelló, P., Blanco, E., Boné, A., & Guerao, G. (2014). Larval stages of the deep-sea lobster *Polycheles typhlops* (Decapoda, Polychelida) identified by DNA analysis: Morphology, systematic, distribution and ecology. *Helgoland Marine Research*, *68*(3), 379–397. https://doi.org/10.1007/s10152-014-0397-0

104. Tshudy, D. (2013). Chapter One—Systematics and Position of Nephrops Among the Lobsters. In M. L. Johnson & M. P. Johnson (Eds.), *Advances in Marine Biology* (pp. 1–25). Academic Press. https://doi.org/10.1016/B978-0-12-410466-2.00001-7

105. Uglem, I., Belchier, M., & Svåsand, T. (2005). Age Determination of European Lobsters (*Homarus Gammarus* L.) by Histological Quantification of Lipofuscin. *Journal of Crustacean Biology*, *25*(1), 95–99. https://doi.org/10.1651/C-2448

106. Wakabayashi, K., Sato, R., Ishii, H., Akiba, T., Nogata, Y., & Tanaka, Y. (2014). Culture of phyllosomas of *Ibacus novemdentatus* (Decapoda: Scyllaridae) in a closed recirculating system using jellyfish as food. *Aquaculture*, *s 330–333*, 162–166. https://doi.org/10.1016/j.aquaculture.2011.12.005

107. Warming pushes lobsters and other species to seek cooler homes. (2019, April 11). Retrieved June 29, 2022, from Science News for Students website: https://www.sciencenewsforstudents.org/article/global-warming-climate-change-lobsters-ecosystems-migration

108. Woodings, L., Murphy, N., Jeffs, A., Suthers, I., Liggins, G., & Strugnell, J. M. (2019). Distribution of Palinuridae and Scyllaridae phyllosoma larvae within the East Australian Current: A climate change hot spot. *Marine and Freshwater Research*, *70*. https://doi.org/10.1071/MF18331

109. Yang, C.-H., Bracken-Grissom, H., Kim, D., Crandall, K. A., & Chan, T.-Y. (2012). Phylogenetic relationships, character evolution and taxonomic implications within the slipper lobsters (Crustacea: Decapoda: Scyllaridae). *Molecular Phylogenetics and Evolution*, *62*(1), 237–250. https://doi.org/10.1016/j.ympev.2011.09.019

110. Yang, C.-H., Lin, C.-W., & Chan, T.-Y. (2014). <p><strong>Additional slipper lobsters of the subfamily Scyllarinae Latreille, 1825 (Crustacea, Achelata, Scyllaridae) from Taiwan</strong></p>. *Zootaxa*, *3852*(3), 336–346. https://doi.org/10.11646/zootaxa.3852.3.3

111. Zhang, C., Li, Z., Li, F., & Xiang, J. (2014). Effects of starvation on survival, growth and development of *Exopalaemon carinicauda* larvae. *Aquaculture Research*, *46*. https://doi.org/10.1111/are.12386

INDEX

C

K

L

M

N

O

P

Q

R

S

FSC
www.fsc.org
MIX
Papier | Fördert gute Waldnutzung
FSC® C083411

Kolibri 360 GmbH
Ferdinand-Jühlke-Straße 7
99095 Erfurt, Deutschland
produktsicherheit@kolibri360.de